CAE Gold PLUS

teacher's book

Norman Whitby

Contents

Introduction

Student profile

The students with whom you will be using this course will have studied English for approximately 700 to 800 hours and will now be planning to take the Cambridge Certificate in Advanced English (CAE). They may already have taken Cambridge First Certificate (FCE) or one or more Certificates in English Language Skills (CELS) at Vantage or Higher level.

The CAE corresponds to the **Council of Europe Framework** level C1. According to this framework, learners of English at this level can function as follows in the language and skills areas described below.

Grammar

Students at this level have a good degree of grammatical control and do not generally make mistakes which lead to misunderstanding. Errors may still be made in more complex structures. They will need to revise areas such as these. At the same time, they will also need to develop their knowledge of certain more advanced grammatical structures.

Vocabulary

Students have good range of vocabulary for common topic areas and are able to use a good variety of expressions to avoid repetition. There may be gaps in their vocabulary when dealing with more specialised topics. They will need to develop their awareness of nuances of meaning and concentrate on making their English sound more authentic and natural by focusing on common collocations and expressions. They should work on expanding their knowledge of word formation, phrasal verbs and idiomatic expressions and should be encouraged to make use of a good monolingual dictionary in order to develop their vocabulary.

Reading

Students at this level have well developed reading skills and can scan for relevant information and skim for the main topic of a text. They can grasp the overall meaning of complex authentic and semi-authentic materials and understand complex opinions or arguments as expressed in serious newspapers, using features such as text structure and referencing to help them.

Writing

C1-level students can produce a variety of texts such as formal and informal letters of various types in a consistent register. They are aware of the conventions for organising and structuring different types of texts such as articles, proposals and reports. They can present arguments, persuade and justify their opinions on abstract topics. In general, they are able to communicate their main message clearly in appropriate language so that the text has the desired effect on the intended reader.

Speaking

Students at this level can communicate effectively in a wide variety of situations and can use both formal and informal language appropriately. They can have extended conversations of a casual nature and discuss abstract topics with a good degree of fluency. They can give clear presentations and contribute effectively to discussions by defending and justifying their point of view, and use effective language to persuade and negotiate with others.

Listening

C1 students can deal confidently with most authentic or semi-authentic listening passages. They are able to pick up nuances of meaning and opinion and follow discussions on abstract topics. They can understand most of what is said in a film or a TV or radio programme, although they may be unfamiliar with some idiomatic or colloquial expressions and may have problems understanding some regional accents.

Preparing for the Certificate in Advanced English exam

A CAE course should consolidate and extend what students already know and train them in the specific techniques and strategies required for the CAE exam. During the course, students should try to work independently at times, using and developing their study skills and strategies for improving their language ability. They should be aware of issues such as collocation and register in order to record vocabulary effectively and be able to use grammar reference material in order to cover any gaps in their grammatical knowledge and build on what is done in the Coursebook.

Features of the *CAE Gold Plus* course

Components of the course

The components of the course include the ***CAE Gold Plus Coursebook***, plus cassettes or CDs, the ***CAE Gold Plus***

The Common European Framework and the *Gold* series

The table below gives a general overview of the Common European Framework levels and the Cambridge ESOL main suite and where the *Going for Gold* and *Gold* series fit into this.

Common European Framework	Guided learning hours from beginner	Cambridge ESOL main suite exams	Cambridge ESOL Certificates in Language Skills (CELS)	*Gold* series
A2	Approx. 180–200	KET (Key English Test)		
B1	Approx. 350–400	PET (Preliminary English Test)	CELS Preliminary	*Going for Gold*
B2	Approx. 500–600	FCE (First Certificate in English)	CELS Vantage	*Going for Gold* *First certificate Gold Plus*
C1	Approx. 700–800	CAE (Certificate in Advanced English)	CELS Higher	*CAE Gold Plus*
C2	Approx. 1,000–2,000	CPE (Certificate of Proficiency in English)		*NEW Proficiency Gold*

Exam maximiser with CDs, the ***CAE Gold Plus CD-ROM*** and this teacher's book.

Supplementary materials

A selection of supplementary materials is also available for extra practice and development of vocabulary, grammar, fluency and exam skills, including:

- *Longman Dictionary of Contemporary English*
- *Longman Exams Dictionary*
- *Longman Language Activator*
- *CAE Practice Tests Plus*
- *Grammar and Vocabulary for Cambridge Advanced and Proficiency*
- *Test your Phrasal Verbs* (Penguin English)
- *Test your Idioms* (Penguin English)

CAE Gold Plus Coursebook

Organisation of the Coursebook

The Coursebook offers progressive preparation for the CAE exam, as well as developing and extending students' competence in the language. Exam-style tasks are introduced from the early stages of the book with graded support being gradually withdrawn as the course progresses.

Each of the 14 units provides an integrated package for all five papers in the CAE exam, as well as grammar and vocabulary development and practice, which are grouped around a common theme. Advice on specific language points or strategies for tackling exam-style tasks is offered in the **Tips** boxes. A key feature of each unit is the **Exam Focus** section which presents the techniques and strategies required for a specific task in the CAE exam and provides exam-level practice.

At the back of the Coursebook you will find a section containing visuals for the **Paper 5 Speaking tasks**, a **Grammar reference**, a **Writing reference** and a **Vocabulary reference**. The **Grammar reference** is a mini-grammar covering all the points dealt with in the units. The **Writing reference** contains model answers for the types of writing which may be tested at CAE. There are also authentic student answers which students can evaluate using the general marking guidelines provided. The **Vocabulary reference** contains a listing of lexical items which are found in the Coursebook, together with definitions and examples.

Recycling and revision

Each unit ends with a review of the language presented in that unit except for units 5, 10 and 14. These are followed by progress tests, which take the form of a complete Paper 3 test. These can be used by the teacher in class as reviews or as tests of the students' command of the language presented in the units.

Grammar

Various different approaches are used for the presentation and practice of grammar points. Use of English tasks in exam format also recycle the grammar that has been presented. The grammar sections are cross-referenced to the **Grammar reference** at the back of the book. The Coursebook also features **Watch Out!** boxes which are designed to pick up on common grammar and vocabulary mistakes made by students.

Vocabulary

A variety of presentation and practice techniques is used in ***CAE Gold Plus***. When reading, students are encouraged to work out the meanings of unknown words for themselves and recognise clues such as affixation or explanations in the text. Ways of recording and learning new words are also emphasised. Students are encouraged to use a monolingual dictionary such as the *Longman Dictionary of Contemporary English*, which gives information about meaning, pronunciation, grammar and collocations.

Particular attention is paid to word formation, which builds students' understanding of how prefixes and suffixes are used, followed by regular practice. This is particularly relevant for Paper 3 part 3.

Reading

Authentic texts from a range of sources are used to develop reading skills and techniques for CAE. Students are encouraged to use the titles and subtitles of the text as well as any non-textual information, such as accompanying photographs, to help them predict the content. Guidance is provided to help them do the task and apply appropriate strategies. Vocabulary and discussion tasks after the reading texts allow students to develop the topic further and to focus on key vocabulary from the text.

Listening

The listening texts are also from a range of sources and the recordings present students with a variety of mild accents. Students are always reminded to read through the task before they listen to help them predict what they might hear, and tips and guidance are often provided to help them complete the task.

Writing

Each unit ends with a writing task of a type found in the CAE exam. The section is cross referenced to the **Writing reference** at the back of the book which provides model answers for each of the text types. In each case students are encouraged to read the task carefully, thinking about the intended reader, and what needs to be included. They are guided towards an understanding of the various conventions of the text type, such as register, layout and typical organisation of ideas. They are then presented with a model answer, which is often used for further language work. Finally, they are given the task of writing a similar text themselves, which can be done either in class or as homework.

Speaking

The grammar, vocabulary and skills sections all provide some opportunity for speaking practice by asking students to respond to the topic or text.
Each unit also contains a section with specific speaking practice for Part 5 of the exam. This presents language for such functions as agreeing and persuading as well as techniques such as how to keep the conversation going.

CAE Gold Plus maximiser

Another major component of the course is the ***CAE Gold Plus maximiser***. Working through the exercises in the **maximiser** will help students to consolidate the language and skills presented in the Coursebook and provide them with further exam-specific practice and preparation.

Each of the 14 units corresponds thematically with the units in the Coursebook. The sections within each unit are cross-referenced to the related Coursebook sections and provide consolidation both of language and of skills work. The grammar and vocabulary sections also recycle material presented in the Coursebook, which is then practised further by means of topic-related exam-style Use of English (Paper 3) tasks. Sections containing exam-style tasks provide information about the exam, plus strategies for tackling each task type, and give students the opportunity to put these into practice.

The ***maximiser*** can be used in class in tandem with the Coursebook as a means of providing further work on specific grammar or vocabulary areas or, alternatively, students can do the exercises and skills practice for homework.

CAE Gold Plus CD-ROM

The CD-ROM provides a variety of exercises to recycle and extend grammar and vocabulary areas presented in the Coursebook. The sections again correspond thematically to the units in the Coursebook and many of the exercises, such as multiple-choice gapfills, are in the style of the CAE exam. The CD-ROM can be used in tandem with the Coursebook to provide further grammar and vocabulary work or it can be used as self-access material.

CAE Gold Plus teacher's book

The ***teacher's book*** provides suggestions on how to use the material in the Coursebook to best advantage. Answers to all the exercises in the Coursebook are found at the end of each section of notes. Recording scripts to all of the listening tasks are also provided. **Teaching tips and ideas** provide suggestions for further activities to practise the material or develop study skills. There is also a section of **photocopiable activities** which provide extra communicative practice in key areas of grammar and vocabulary from the Coursebook units. Many of these are directly related to exam-style tasks. Detailed teaching notes state the aims and rationale of each photocopiable activity and provide a step-by-step procedure for using them in class.

You will also find a bank of 14 **photocopiable tests** made up of 11 unit tests and 3 progress tests. The unit tests are based on the language covered in a single unit and should take no more than 30 minutes to complete. The progress tests are to be used after your students have completed units 5, 10 and 14 and should take between 50 and 60 minutes to complete. They revise and test the language covered in the previous four or five units.

UNIT 1 Tuning in

Listening: multiple choice (Part 1) p.6

Aims:
- **to give practice in listening to identify opinion, attitude and general gist**
- **to complete an exam-style listening task (Paper 4, Part 1)**

Exam information

In Paper 4, Part 1, candidates listen to three short extracts and answer two multiple-choice questions on each. Some of the questions focus on the speakers' opinions or feelings.

1 Use one or more of these questions to conduct a brief class discussion on the topic of music. You could personalise the topic by asking if anyone plays a musical instrument or has ever attended a concert.

2 Students read the multiple-choice questions for the first extract. They may find it useful to underline important words in the alternatives. Then do the listening exercise. They compare in pairs before listening again. Follow the same procedure for the other extracts before checking the answers with the whole class.

3 In pairs students compare the types of music that they like or dislike in different situations. You could introduce question 3 by giving examples of people who have made their fortune through singing.

▶ Recording script p.90

ANSWERS

Ex. 2

1 B 2 C 3 A 4 B 5 A 6 C

Grammar 1: overview p.7

Aims:
- **to identify problematical areas of grammar**
- **to raise students' awareness of how they can improve grammatical accuracy**

1 Students work individually for about five minutes to correct the mistakes in the letter. There will probably be some items that they can correct immediately and others which they feel to be wrong but are not able to confidently correct. For these items you can allow them to underline without correcting.

2 If students do not have access to grammar books, you may choose to use Exercise 1 as a diagnostic exercise for yourself to identify areas for which you may need to do remedial grammar work.

3 These questions can be discussed with the whole class. You can also talk about how they like their written work to be corrected (e.g. correction codes, checking each other's work, etc.).
The grammar checklist suggestion should be introduced at the end of the discussion.

ANSWERS

Ex. 1

Hi Carlos

Just touching base to tell (0) ~~to~~ you about the film I went to see last night as you asked. My advice to you (1) ~~are~~ *is* – don't bother with it at all! It was complete rubbish, and a waste of time and money. I really wish I had not gone myself, and if I'd (2) ~~have~~ read the reviews, I'd have given it a miss. I've been going to the cinema regularly (3) ~~since~~ *for* at least six years, and that was by far the worst film I (4) ~~had~~ *have* seen up to now – it's (5) ~~a~~ such *a* terrible film I can't understand how or why they decided to make it. (6) Apart ~~of~~ *from* everything else, I was so bored! So in you might consider (7) ~~to go~~ *going*, you know my opinion now!
Anyway – enough of my complaints – and in spite of my disappointment with this particular film I haven't actually gone off films in general! So on a different topic – I know that you are (8) interested

~~for~~ *in* live music gigs, and I wondered whether (9) ~~might you~~ *you might* like to come with me to the open-air concert in the park next Saturday? It'll be great, and all the others are going. Let me (10) know ~~it~~ what you think – but unless I hear from you by Friday I'll assume you can't make it. I'm attaching some information about the concert with this email so that you can see who is playing, and we can get the tickets on the night.
So that's all for now – speak to you soon.
All the best,
Jose

Reading: multiple matching (Part 4) p.8

Aims:

- **to practise gist reading to identify the focus of each section**
- **to complete an exam-style reading task (Paper 1, Part 4)**
- **to give practice in inferring word type and meaning from context**
- **to use a dictionary to find example sentences and collocations**

1, **2** Write the term *tribute band* on the board and ask students if they know what it means. Then ask the class's opinion on the three gist questions. Students then read the text quickly to find the answers. When checking the answers, ask students which section of the text they found each answer in and ask them to summarise the topic of each section. For example, B deals with who goes to see tribute bands.

3 Students first read questions 1–15. Tell them that it may help if they underline the important words in each question, such as *preparation* and *one tribute band* for question 1. Ask if they know any answers from the initial gist reading. Then students complete the reading task, with a time limit of about 15 minutes. They should read each question and then search for the corresponding reference. If they cannot find it, they should move on and come back to that question at the end.
After 15 minutes students compare their answers in pairs before checking as a whole-class activity.
(A more detailed procedure for Paper 1, Part 4 is given in Unit 5 of the Coursebook.)

4 Students scan the text and underline any 'copying' words or phrases. Then give a dictionary to each pair or group and ask them to check the meanings of any words which were new. This is an opportunity to point out the kinds of information which a dictionary provides, such as example sentences.

5, **6** Students look back at the text and underline the words in the list. They then decide the type of word and the meaning. Point out that they can use both affixation (an obvious example here is the *-ing* ending) and context to infer word type. Then students turn to page 188 to check with the dictionary entries.

7 This is an opportunity to hold a class discussion on the use of dictionaries.

ANSWERS

Ex. 3
1 B **2** E **3** B **4** C **5** D **6** E **7** A **8** C **9** E
10 AB **11** AB **12** BC **13** BC **14** DE **15** DE

Ex. 4
a musical impression covers pop faker
facsimilies copycat ersatz clones
impersonating impersonators imitators

Vocabulary: word formation (suffixes) p.10

Aims:

- **to revise and extend students' knowledge of suffixes**
- **to provide practice for Paper 3, Part 3**
- **to practise an exam-style sentence transformation**

1 Students identify the part of speech. Ask them to give other examples of adjectives and verbs which can take these suffixes.

2 Students now work in pairs to identify the word types associated with each suffix.

3 Use this question to check students' answers as a whole-class activity.

4 This extends the exercise by asking students to give examples.

Teaching tips and ideas

Students should be encouraged to record suffixes as part of their vocabulary notes. One way of recording these is to make word diagrams like this:

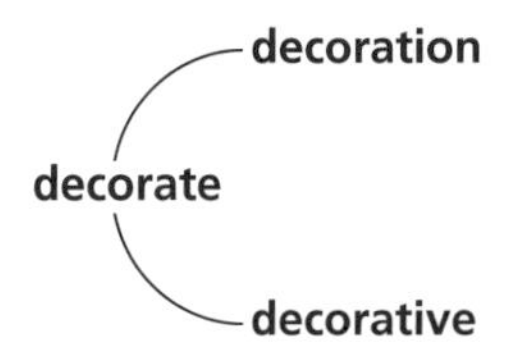

5

1, 2 Students read the title of the article and speculate about the content before skimming the article.
3 After this initial reading, students work in pairs to put the words in brackets into the correct form.

6 After checking the answers, use this question to personalise the topic by applying it to the students' own town or city, or if there are no buskers, you might like to ask students how they feel about Peter Murphy's decision to give up a steady job as an accountant to become a busker.

7 Point out that the adjective *disappointed* in the example needs to change to *disappointment* in the new sentence. Then ask students to complete the four transformations.

8 These questions personalise the topic of the reading text.

ANSWERS

Ex. 1
All the words are nouns, because the suffixes *-ment* and *-ness* are noun suffixes.

Ex. 2
weak**ness** (noun), count**able** (adjective), alterna**tive** (noun/adjective), frighten**ing** (adjective), rapid**ly** (adverb), op**tion** (noun), success**ful** (adjective), perform**ance** (noun), leg**al** (adjective), responsibil**ity** (noun), modern**ise** (verb), politi**cal** (adjective), enjoy**ment** (noun), delic**ious** (adjective), decora**tive** (adjective), confus**ed** (adjective), pleas**ant** (adjective), combin**ation** (noun)

Ex. 3
nouns from adjectives: -ity, -ness
nouns from verbs: -ive, -ion, -ment, -ance
verbs from adjectives: -ise (NB US spelling -ize)
adjectives from nouns/verbs: -ive, -able, -ed, -ing, -al, -ed, -ant
adverbs from adjectives: -ly

Ex. 5
1 professional **2** sponsored **3** regularly
4 determined **5** impression **6** intriguing
7 production **8** inspirational/inspiring **9** formal
10 powerful **11** fluently **12** respectable
13 appearance **14** responsibility
15 involvement **16** negotiations
17 determination **18** performances

Ex. 7
1 much more forgetful **2** be more responsible
3 a brilliant performance **4** was actively engaged

▶ Photocopiable activity 1 *Suffixes* pp.154 and 155

Use of English: open cloze (Part 2) p.12

Aim:
- **to complete an exam-style open cloze**

1 If you have already discussed students' opinions about different types of music, just use question 2.

2 Students work in pairs to list the advantages and disadvantages and then tell the class their ideas. Then they skim the text to see which of their ideas are mentioned.

3 Go over the procedure outlined and then ask students to work individually to complete the task. They then compare answers in pairs and guess the words for any remaining gaps (stages 2 and 3). Step 4, re-reading the whole text, is important to ensure that students' answers fit with the overall argument. You could set aside a special minute for this.

4 The first question checks students' understanding of the overall opinion.

ANSWERS

Ex. 2
1 whose **2** no **3** from **4** on **5** how
6 what **7** other **8** in **9** to **10** however
11 gave/give **12** nobody/no(-)one
13 rather **14** as **15** which

Exam focus
Paper 5 Speaking: conversation (Part 1) p.13

Aims:
- **to provide students with information about Paper 5 (speaking), Part 1 and allow them to practise**
- **to help students to analyse what makes a good candidate**

1 Go over the exam information with the students. Then play the recording and invite students' comments on the candidates.

2

1 Students now work in groups of three to do mock interviews. In the discussion afterwards encourage students to think about what could have been said to make their answers more detailed or interesting. If some candidates try to say too much, you may need to tell them that just two or three sentences will do at this stage.
2 If students find it difficult to think of questions, prompt them by writing possible topics on the board (e.g. *TV*, *weekends*, etc.).

▶ Recording script p.90

ANSWERS

Ex. 1
Brita needs to be more imaginative and explain her reasons. She hesitates, and should try to be more fluent.
Petra gives interesting details. She uses good interactive language – *I agree with you, you know,* etc. She picks up on what Brita has said.

Grammar 2: verb tenses (perfect aspect) p.13

Aim:

- **to revise and extend students' knowledge of perfect tenses and the distinction between simple and continuous**

1 Students look at the example sentence and identify the order of events.

2

1, 2 Do these with the whole class to check familiarity with perfect tense forms. Point out that the use of *by* in the sense of *before or no later than* is often associated with a past perfect or future perfect tense.
3 After correcting the mistakes, students should read out the correct versions pronouncing the contractions.

3

1 Students work individually before checking in pairs.
2 This can be done as a whole-class discussion.

4 Students work in pairs to discuss the differences between the sentences. Emphasise the difference in particular between sentences in pairs 2 and 6, where the use of the wrong tense could cause misunderstanding.

5 Students now work in pairs to complete the exercise.

6 This discussion activity gives students an opportunity to use perfect tenses in a freer context. Give ten minutes for students to find something true for both of them for each question. Then ask pairs to tell the class their most interesting example.

ANSWERS

Ex. 1
first event = past perfect – *had downloaded the songs*
second event = past simple – *realised how good they were and forwarded the files*

Ex. 2
1
1 's been 2 'll have finished 3 'd seen
4 've been
2
1 present perfect a) 2 future perfect d)
3 past perfect c) 4 present perfect b)
3
1 After I'd been there (past simple *vs* past perfect)
2 she'll've been away (future simple *vs* future perfect)
3 Jose went to the football game (past simple *vs* present simple)
4 He has always enjoyed (present simple *vs* present perfect – <u>state</u>)
5 I'll've finished (present simple *vs* future perfect)
6 She has visited (present simple *vs* present perfect – <u>event</u>)

Ex. 3
1
B 2 'd been staying
C 3 'll have been working
D 3 's been practising
E 3 'll have been waiting
2

Results apparent later	D
Temporary activity or state	B
Duration emphasised	E
Incomplete action	C

Ex. 4
1 a) we focus on the result, b) we focus on the activity itself.
2 a) is a present temporary situation, b) is a temporary situation which may or may not still be the case.
3 a) and b) are very similar, but a) focuses more on the duration.
4 similar, but (b) has temporary implications.
5 no difference.
6 a) in my life so far b) during a specific time in the past.

Ex. 5
1 've been listening **2** went **3** hadn't been
4 will have learned **5** has ruined
6 will have been playing **7** were standing
8 had been waiting

Writing: drafting and organising (Part 1) p.15

Aims:

- **to raise students' awareness of the skills involved in producing a written text**
- **to raise students' awareness of how their writing will be assessed in CAE Paper 2**
- **to complete an exam-style writing exercise (Paper 2, Part 1)**

1 Students read the five statements and discuss in pairs whether they think they are true or not. You could extend the discussion by asking students to reflect on what stage of the writing process they focus on. As a general rule, they should consider the audience, and plan and connect their ideas first and leave checking grammar and spelling until later.

2 Students do the matching exercise.

3

1. Students discuss the three questions briefly in pairs.
2. Students read the task carefully to themselves and underline the three points that the answer must deal with.

4

1. This is best done as a whole-class activity. Most students at this level should be familiar with the conventions for formal letters.
2. Students read the model letter on p.191 and check it against each question on the list.

5 Students work in pairs to complete the plan of the letter. Go over it with the whole class, asking them to suggest other linking words which could be used.

6 Students read the new task, underline the three areas to be covered and decide on the most logical order. Point out that paragraphing will reflect this.

7 This can be done in class or for homework.

8 If the writing task was done for homework, this activity can be done the following lesson. Students should hand in their letters only after they have been evaluated with the checklist. Some students may wish to write an improved version in response to the evaluation, which they can give in later.

ANSWERS

Ex. 2
a) 2 b) 3 c) 4 d) 5 e) 1

Ex. 3
2
outlining the reasons why you were disappointed
your cousin's reaction to the film
suggesting what should be done

Ex. 4
1
1 reason for writing
2 explanation/clarification of the situation
3 further supporting details
4 any requests for action, or further information
2
2 Yes, but has expanded on some. It is not always necessary to use every point but the writer should choose the most appropriate points to answer the task.
3 Practical problems, inappropriateness of film
4 Yes
5 a) uncomfortable seats, single seller, missed bus
b) cousin unable to sleep
6 Semi-formal
7 No – to complete the task fully and appropriately it is necessary to add more weight to some points.

Ex. 5
Opening paragraph: Reason for writing
Information included: background situation
Linking phrases: *I am writing to ...*

Second paragraph: *Practical* problems
Information included: *started late/no ice cream/ missed bus/uncomfortable seats*
Linking phrases: *Firstly, so, although, On top of that*

Third paragraph: More *suitability* problems
Information included: *Inaccuracy of advertisement, emotional problems*
Linking phrases: *However, Despite the fact that, In fact*

Final paragraph: *Suggestion and solution*
Information included: *possible future action*
Linking phrases: *I suggest that, thus*

UNIT 1 Review p.17

ANSWERS

Ex. 1

1 In the first place **2** even though **3** Secondly
4 in spite of **5** Furthermore **6** Finally **7** as

Ex. 2

1 We *never go* out ...
2 Where are *the* tickets?
3 ... whether *there is* another cinema ...
4 The new arts centre *is* very nice ...
5 That's the boy *whose* brother ...
6 The tourist board gave us lots of *information* ...
7 Unless you *work* harder... OR *If* you don't work harder ...
8 We considered *going* to the concert ...
9 I really wish I *had* more time to study!
10 She apologised for *being* late.
11 I *know* she enjoys ...
12 I can't get used to *starting* ...
13 ... despite ~~of~~ her fear of heights.
14 My teacher won't let me ~~to~~ get out of doing homework.
15 If I'd ~~have~~ known ...
16 They've been living in this town *for* at least 25 years.
17 He's such *a* hard worker ...
18 Computer games are a lot ~~more~~ cheaper now ...
19 I found the film absolutely *terrifying*.
20 He might *decide* ...

Ex. 3

1 enjoyable **2** standardise **3** donation
4 hopeful **5** financially **6** productive
7 disappointment **8** weakness **9** outrageous
10 acceptable **11** frightening **12** uplifting

UNIT 2 Spend it or save it

Listening 1 p.18

Aims:
- **to give practice in listening for specific information**
- **to review / introduce expressions to do with saving and spending**

1

1 This can be used as a whole-class activity to introduce the theme of spending money. If you are teaching in a country which does not use the euro, give similar amounts of the students' own national currency.
2 In pairs or groups, students think of ways in which they could save money. They then report back to the class.
3 Discuss these questions briefly with the whole class.

2 Play the first part of the recording and ask students as a whole class to explain the answers to 1 and 2.

3

1 Ask students to look at the categories of people and tell you what they think the words mean. You will probably need to teach the verb *scrimp* and the collocation *scrimp and save*. Play the recording so that students compare their ideas with the psychologists' descriptions. Then ask them to match statements A to F to the type of spender.
2 This is a chance to personalise the topic of the listening task.

4 Students now listen for the specific advice. After listening, they compare notes in pairs.

5 Students can work in pairs to divide the expressions into the two groups. After they have completed this, play the recording again, pausing after the description of each kind of person. Ask which expressions they heard in each section and check the answers to the vocabulary exercise.
As a possible follow-up activity, students could work in pairs to personalise this topic. Give them some suggestions (e.g. they could talk about a time they went on a shopping spree, made a sound investment, bought something on impulse, ran out of cash, gave themselves a treat, etc.).

▶ Recording script p.90

ANSWERS

Ex. 2
1 How saving a little each day can make a difference
2 There's more to life than saving

Ex. 3
1 A C
2 D E
3 B F

Ex. 4
1 pay bills online
2 pay for things by cheque
3 set a budget that includes treats

Ex. 5
a) interest, set a budget, a sound investment, a nest egg, to economise, put it away for a rainy day
b) conspicuous consumption, go on a spree, run out of cash, a treat, in the red, shopaholic, get through money like water, on impulse, a 'must-have' item

Speaking: giving opinions p.19

Aim:
- **to provide practice in speaking for Paper 5, Part 3**

1 Play the recording and ask students to summarise what the candidates have to do.

2 Students sometimes make the mistake of describing the pictures instead of discussing the given issues, and so the first question is intended to pre-empt this. After they have identified the agreeing and disagreeing phrases, ask them to suggest others. Students often overuse *I agree* whereas native speakers prefer other phrases such as *absolutely*. Watch out for the common error *I am agree*.

3 Students now do the speaking task in pairs. Encourage them to use a range of expressions for giving opinions, agreeing or disagreeing.

4 Discuss this with the class. You could compare these with other non-material things that are important, such as health.

▶ Recording script p.91

ANSWERS

Ex. 1
Explain why the things have become important, decide which two are not necessary

Ex. 2
1 no, because they have to discuss the ideas behind the pictures not describe them. They have to explain why these things have become important.
2 Give their own opinion: *Well, yes, actually I do think that; I really believe ...; that's what I'm saying; it still seems to me ...*
Ask for their partner's opinion: *Do you think ...; but don't you think that ...; Why do you think ...;*
Agree with their partner: *Yes, that's right; Yes, all right – you've got a point there; Absolutely*
Disagree with their partner: *I'm not entirely sure that I agree with you there; I just don't accept ... ; Well, even if I go along with that ...*

Exam focus

Paper 1 Reading: multiple choice (Part 1) p.20

Aims:
- **to provide an introduction to the new Paper 1, Part 1**
- **to give practice in answering multiple-choice questions**

Exam information

In CAE Paper 1, Part 1, there are three texts with six multiple-choice questions. The texts have a common theme but may come from different sources and display different purposes and opinions.

Go over the exam information section and suggested procedure. Ask students to suggest any other tips for answering multiple-choice questions.

1

1, 2 Ask students to read the first text quickly, giving them a time limit of about 30 seconds. Then ask them to read the two multiple-choice questions carefully.
For question 1, ask them to scan the paragraph for the words *problem* and *job* to locate the relevant part of the paragraph.
When going over the correct answers to any multiple-choice question, it is useful to discuss why the other alternatives are wrong. For example, in question 1, A and B are incorrect because we learn that she is *known to be very good at her job*, and D is incorrect because we learn that she has *a stylish dress sense*.

3, 4 Follow the same reading procedure as with the first text. Establish that question 3 deals with the writer's purpose in the text and that question 4 is asking for the meaning of a specific phrase. Ask what phrase in question 4 they could scan for to locate the correct section of the paragraph (*immense satisfaction*) and how they know that it appears in the text (it is in inverted commas). Then students work individually to choose the correct answers.

5, 6 Follow the same reading procedure as for the previous two texts. Establish that question 5 again deals with the exact meaning or implication of a phrase and that question 6 deals with the reference system of the text in that it requires students to understand what previous idea *it* refers to.

2 Discuss these opinions briefly with the whole class.

ANSWERS

Ex. 1
1 C 2 D 3 C 4 C 5 A 6 B

Grammar 1: defining and non-defining relative clauses p.22

Aim:
- **to review the grammar of defining and non-defining relative clauses, and the use of relative pronouns**

1

1 Students may already be familiar with the basic distinction between defining and non-defining clauses, but may still have difficulty distinguishing between the two and feel uncertain about when to put a comma. Go over the first example with the class and then elicit the differences between the other pairs. In 2, ask them in which sentence there was more than one charity (the second, as the defining clause here must indicate that there was one charity which the man preferred and another or others he did not). In 3, ask them how many sisters the speaker has.
2 Complete the rules as a whole-class activity. At this point you could check students understand the use of *whom*. *Whom* is not very often used in modern English; it is, however, still used after a preposition. Tell students that in spoken English it is more usual to say *That's the woman who I gave a lift to.*

2 Students rewrite the sentences individually and then elicit the rule.

3 Students work individually to transform the sentences.

4

1 Write the words *charity* and *celebrity* on the board and ask students to give some examples of each. Ask them if they know any celebrities who are involved with a particular charity (Bob Geldof might be a well-known example). Then they skim the text and answer the gist questions.

2 Students work in pairs to complete the gaps.

5 This exercise gives oral practice in using relative pronouns. Do an example with the whole class first by choosing one of the categories and giving a definition so that they can guess the word. They then do the activity in pairs or small groups.

6 If students have little experience of charity events, you could change the discussion into a simulation by telling them that they have been given the task of raising a certain amount of money for a charity that they know, and have to plan how they are going to do it.

ANSWERS

Ex. 1

1

1 a) defining b) non-defining
2 a) non-defining b) defining
3 a) defining b) non-defining

2

1 non-defining
2 who for people and which for things
3 defining
4 Whose

Ex. 2

1 The school where I first studied Economics was in London.
2 Wednesday is the day of the week when I always have a meeting.
3 It was a conference in Rome where I met my future husband.
4 Two o'clock is the time when I always have a cup of coffee.

With relative clauses of place and time, use *where* or *when* instead of *at which* or *on which*.

Ex. 3

1 She gave me her email address, which was how we managed to contact her later.
2 I spent the money on a new car, which was what I had always planned to do.
3 Her face was red, which was how we knew she was upset.
4 He left at six, which was when she arrived.
5 The actor forgot his words, which was why they brought the curtain down.
6 I had a holiday in Spain, which was where I learned to swim.

Ex. 4

1

a) image and career enhancement
b) people will remember the charity and support it themselves

2

1 which/that 2 who/that 3 which
4 which/that 5 which 6 who/that 7 that
8 where/when 9 whose 10 whom

Use of English: word formation (Part 3) p.23

Aim:

- **to complete an exam-style word formation exercise (Paper 3, Part 4)**

1 Write the word *auction* on the board and check that students understand it. If no one has experience of buying or selling anything in this way, they can simply suggest possible advantages and disadvantages.

2 Ask students to skim the text quickly and answer the gist questions. They then complete the word building exercise.

3 Discuss this with the class.

ANSWERS

Ex. 2

1 a) easy to buy b) hard to sell
online facilitator

2

1 possessions 2 unwanted 3 pleasure
4 frankly 5 solution 6 remarkably 7 variety
8 potential 9 percentage 10 commission

Vocabulary 1: compound adjectives p.24

Aim:

- **to introduce or review compound adjectives**

Teaching tips and ideas

The exercises in this section provide a good opportunity to point out the usefulness in general of students recording full collocations in their vocabulary notes. Pages in the notes can be set aside for common collocations around a key word or theme. These can be added to as an ongoing activity. This technique of recording vocabulary also helps students to prepare for Paper 3, Part 5 (gapped sentences).

1 Students work in pairs to match the words to make compound adjectives. Encourage students to guess any that they are unfamiliar with before using a dictionary. They then decide how each adjective might be used.

2 This can be done as a whole-class activity.

3 For this exercise, students focus first on the collocation and guess the meaning if it is not already known. Then, they read the last part of the sentence and suggest a correct alternative.

4 Students now listen to the recording and match each speaker to the correct summary.

5

1 Students now choose the correct prepositions in pairs before checking as a whole class or with a dictionary.
2 This is a brief follow-up to Exercise 5.1, and aims to help students to remember the adjectives through personalisation.

▶ Photocopiable activity 2A *Compound adjective snap* p.156

▶ Recording script p.92

ANSWERS

Ex.1

1+2

old-fashioned (person or thing)
self-centred (person)
last-minute (plan or idea)
far-fetched (idea)
air-tight (thing)
long/short-term (plan)
level-headed (person)
quick-witted (person)
so-called (person or thing, e.g. *expert*)
long-standing (plan, e.g. *agreement*)
mass-produced (thing)
self-made (person, e.g. *millionaire*)

Ex. 2

1 last-minute **2** level-headed **3** far-fetched
4 self-centred

Ex. 3

1 we decided a long time ago.
2 it's just the same as all the others.
3 her no time at all to think of a reply.
4 so it stops your food going dry in the air.
5 he started with no financial help at all.

Ex. 4

1 = speaker 5 **2** = speaker 2 **3** = speaker 1
4 = speaker 3 **5** = speaker 4

Ex. 5

1 hard up **2** run-down **3** one-off **4** worn out
5 well-off **6** burnt out **7** fed up

Listening 2: multiple choice (Part 3) p.24

Aims:

- **to give practice in understanding the speakers' attitude and opinion**
- **to give practice in answering multiple-choice listening questions for Paper 4, Part 3**

1

1 This is a lead-in to the listening activity for students to discuss in pairs.
2 This can be discussed as a whole-class activity.

2 The multiple-choice questions focus on the speakers' opinions. In order to answer them successfully, students need to understand the speakers' overall argument, not specific information. After students have read the questions, point this out to them, and warn them against basing their answers on a single word or phrase. For example, the phrase *I think this is very worrying* in Graham's first utterance may lead students to incorrect alternative C just because of the similar phrase *feels concerned.*
You can encourage students to follow the overall argument by asking them to focus on the links between the ideas in individual questions. For example, question 6 asks about a cause and effect.
At the end, play the recording again to check each answer.

3 This can be kept as a brief whole-class discussion.

4 Students can discuss this question in pairs. If they do not know the same people, they could write down what they have decided to buy and where and then explain their choice to their partner.

▶ Recording script p.92

ANSWERS

Ex. 2

1 D **2** C **3** A **4** D **5** A **6** B

Vocabulary 2: advertising and marketing p.26

Aim:

- **to introduce or review further verb–noun collocations and compound adjectives**

1

1 Introduce the topic by giving an example of an advertisement that you feel is successful. Then students talk in pairs. In a multinational class, they can compare advertisements in different countries.
2, 3 These points are best discussed briefly with the whole class.

2

1 Ask students to read the text quickly, ignoring the gaps, and answer the gist question.
2 Students now work in pairs to fill the gaps before checking the answers as a whole class. Remind them that the answers depend on collocation.
3 Students underline the collocations in the text or record them in their vocabulary notebooks. Ask them to suggest other collocations for these nouns (e.g. *attract someone's attention*).

3

1 Students brainstorm all the places where they can see advertisements. If the words *hoarding* and *flyer* do not come up in the brainstorming, pre-teach them before students read the text.
2 Students work in pairs to complete the compound words.

4 Students discuss these questions in pairs or small groups and then report their opinions.

Teaching tips and ideas

The activity of thinking of their favourite advertisements may not be suitable in a multinational class, where students will know different ones. In this case, you could bring some advertisements into class, give one to each pair of students and ask them to comment on the techniques, the type of consumer targeted and how effective the advert is.

ANSWERS

Ex. 2

1 To involve the reader immediately and reinforce the message.

2

1 adopt 2 create 3 grab 4 bring 5 finishes 6 drives

3 adopt an approach; create an image; grab someone's attention; bring something to mind; drive a message deep

Ex. 3

2

1 pop-ups 2 mass-market 3 high-profile 4 highly regarded 5 so-called 6 identifiable 7 real-life

Grammar 2: articles p.27

Aims:

- **to review the grammar of articles**
- **to give practice for Paper 3, Part 3**

1 Students work in groups to brainstorm brand names, possibly taking two or three items each.

2 Ask students to read the text quickly, ignoring the gaps, and answer the gist question.

3 Students now work in pairs or individually to complete the gap fill. Articles can be a very problematic area, especially for students whose native language may not have them.

4

1, 2 Students do these exercises individually at first and then compare answers.
3 Briefly discuss students' reaction to the two articles.

▶ Photocopiable activity 2B *Advertising techniques* p.157

Teaching tips and ideas

To extend this topic, ask students to work in groups to design a logo and invent a slogan for a given product. Give them a choice of three (e.g. trainers, toothpaste, fruit juice). Provide each group with an OHT or paper to make a poster so that they can draw the logo that they decide on and present it to the rest of the group. One person from each group should be chosen to talk for approximately one minute (as in CAE Paper 5) and then invite questions.

ANSWERS

Ex. 2

b)

Ex. 3

1 What is a brand? **2** a car **3** the brand name **4** the design or packaging **5** the special features of **6** the world **7** the consumer **8** brand names **9** school

Ex. 4
1
1 a brand name 2 insurance (no article)
3 a group
4 a desirable lifestyle 5 the world 6 a brand
7 the brand image
2
1 an advertisement 2 a car
3 the company's advertisement
4 the poor quality 5 the determining factor
6 people (no article) 7 a strong response
8 a product

Writing: informal letter (Part 2) p.28

Aim:

- **to complete an exam-style writing question (Paper 2, Part 2) requiring students to produce an informal letter**

Exam information

In CAE Paper 2, candidates are required to answer one compulsory question and choose a second question from four alternatives. The compulsory question can be on a number of different genres including a letter, report or article but the task will always involve persuasion in some form.

1

1 Students read the task and identify which part of it explains the situation (the first part) and which tells them what they have to do (the second). Then they read the second part and underline the relevant phrases.
2 Students talk in pairs and then tell the class their ideas.

2

1 Students read the letter and answer questions 1 and 2 as a whole-class activity.
2 Students work alone or in pairs to find and underline the expressions. Some of them are collocations which could be recorded in the students' vocabulary notes (e.g. *have a go*).

3 Students decide in pairs how the last two paragraphs should change and then write their own improved version.

4 This writing could be set for homework, but it may be better to do it in class if the group are relatively unpractised at CAE writing tasks.

ANSWERS

Ex. 1
explaining what happened, what you did about it, how it affected you and advising your friend

Ex. 2
1
2 no – they haven't advised their friend
2
1 had a go 2 a bit of a disaster 3 a downside
4 home and dry 5 watch their backs
6 at a loss to know what to do

UNIT 2 Review p.29

ANSWERS

Ex. 1
1 unharmed **2** tendency **3** exposure
4 unacceptable **5** outlets **6** examination
7 replacement **8** applications **9** unfortunate
10 equally

Ex. 2
1 in a house which has (got)
2 it very difficult to live without
3 (her) support to a number of
4 (which) I like best is (the)

Ex. 3
1 a **2** a / the **3** the **4** ~~a / the~~ **5** ~~a / the~~
6 the **7** the **8** a **9** the **10** the **11** ~~a / the~~
12 the **13** ~~a / the~~ **14** a **15** the

UNIT

3 What makes us tick

Vocabulary: adjectives of character p.30

Aim:

- **to extend students' knowledge of personality adjectives and idioms to describe personalities**

1 Write the term *reality TV* on the board and ask what students understand by it (*real people, not actors, in real situations*) and if they know any examples. The 'Big Brother' format has been televised in many countries and so many students will probably have heard of this. Then go on to ask the questions in the book about how people are chosen and why.

2

1 Students read the profiles and underline the personality adjectives. You could tell them to double underline any adjectives for which they are not sure of the meaning. Then they can work in pairs to compare which adjectives they knew and explain the meanings if necessary.
2 Students talk in pairs to choose one adjective from each profile. At the end, ask the class which adjective was most commonly chosen for each person.

Watch Out! *sensible/sensitive; sympathise/empathise*
Sensible is a well-known false friend, as many European languages have a similar word which means *sensitive*. The second pair of sentences highlights the difference between *sympathise* and *empathise*. Again this problem is often compounded by the existence of a false friend. Many European languages have a word similar to *sympathetic* which simply means that you get on well with that person.

3 Students discuss briefly in pairs or groups who they think has the best reason for wanting to take part. This should lead naturally into the discussion in task 2 where students select five personalities. At the end, groups report their decision to the whole class.

ANSWERS

Ex. 2

2

Alain: idealistic, conscientious, well-organised
Cris: ambitious
David: quiet, sensitive, self-conscious
Ella: curious, independent, taciturn
Franz: trustworthy, supportive, reliable, serious
Gina: playful, high-spirited, undisciplined, impatient, extrovert
Harold: quick-tempered, assertive, self-opinionated
Iva: normal, sociable, not confrontational
Brita: caring, empathetic, sincere, warm-hearted, sentimental
Positive: idealistic, conscientious, well-organised, sensitive, curious, independent, trustworthy, supportive, reliable, playful, high-spirited, assertive, sociable, caring, empathetic, sincere, warm-hearted
Negative: self-conscious, taciturn, undisciplined, impatient, quick-tempered, self-opinionated, confrontational, sentimental
Either: ambitious, quiet, extrovert, normal

Watch out!

1 a) sensible b) sensitive
2 a) sympathise b) empathise

Grammar 1: modal verbs 1 p.31

Aims:

- **to revise common modal verbs and clarify students' knowledge of their meanings**
- **to give further practice with modal verbs in the context of an exam-style sentence transformation activity (Paper 3, Part 5)**

1 Students complete the matching exercise and then compare answers in pairs.

2 Ask students to work individually to complete the transformations and then compare their answers in pairs. Then go through the answers with the whole class, pointing out how the modal meanings are expressed in different ways such as *is compulsory* for *has to*.

3

1 Students read the advice and complete the gap-fill exercise before comparing in pairs.
2 Students now work in pairs to write an additional two or three sentences. Pairs then read their advice to each other.
3 Begin this speaking activity by telling the class about an example of your own and then asking them to talk in

pairs. You could allow them to choose just one of these situations if they prefer.

ANSWERS

Ex. 1
1 e) **2** c) **3** a) **4** f) **5** d) **6** b) **7** g) **8** h)

Ex. 2
1 didn't have to take
2 could/may/might find this book helpful
3 chances are (that) he'll
4 that/it must be him
5 has to work

Ex. 3
1 ought to **2** might **3** can **4** have to **5** can't
6 could **7** may **8** must **9** shouldn't
10 don't have to **11** mustn't **12** can

▶ Photocopiable activity 3 *Personality types* pp.158 and 159

Exam focus

Paper 4 Listening: multiple matching (Part 4) p.33

Aims:

- **to give practice in listening to identify attitudes**
- **to complete an exam-style listening task (Paper 4, Part 4)**

Exam information

In Paper 4 (listening), Part 4, students listen to five extracts. There are two sets of questions, both involving matching. Students should focus on the first set of questions on the first listening and the second set when the extracts are repeated. The questions focus on attitude, opinions and context rather than specific information.

1 Go over the exam information and suggested procedure with students. Then ask them to read the two tasks and underline the most important words in both the main question and the alternatives. Point out that general or 'vague' information in the options is likely to be more specific in the actual recording. For example, if option C is used, the recording is likely to name a specific person who could not pronounce the name.
Students listen to the recording for the first time and do task 1. They compare their ideas in pairs before listening again and focusing on task 2.
When going over the answers, play the recording again, pausing after the key sentence in each extract such as *I really felt that my name stopped me from standing out in a crowd* for Speaker one. Point out how sometimes students can eliminate some answers before they hear the correct one. For example, the phrase *I wasn't made fun of or anything* for Speaker three eliminates option F before students hear the correct answer.

▶ Recording script p.93

2 Students now match the phrases from the recording to the closest meaning. Check answers as a whole class.

3 This discussion allows students to personalise the topic of changing names from the listening test.

ANSWERS

Ex. 1
1 H **2** A **3** C **4** E **5** D **6** H **7** E **8** C
9 G **10** D

Ex. 2
to laugh it off = not to take too seriously
to stand out in a crowd = be distinctive
really fed up = very unhappy
I happened to = by chance
to get his tongue round = pronounce
to split up = end a relationship
made redundant = lost a job
did the trick = achieved its aim
a snap decision = happened quickly
to tease = to make fun of

Reading: multiple choice (Part 3) p.34

Aims:

- **to introduce some ways of apologising in English**
- **to give practice in identifying opinions and how they are supported in the text by reference to other authorities**
- **to complete an exam-style multiple-choice exercise**

Exam information

In Paper 1, Part 3, candidates answer seven multiple-choice questions on a text. The questions can test understanding of both specific details and the writer's overall opinion. Sometimes the question may explicitly direct students to a particular paragraph; if not, they should try to pick out a word in the stem which they can look for in the text to help them locate the answer.

1 Students read the two sayings and then comment. You might develop the discussion of the second saying by asking if loving someone means that you never hurt them.

2 Write *I'm sorry* on the board and elicit some adverbs which could be used to make the apology stronger (e.g. *really, terribly, awfully*). Then ask students if they know any other formulas which could be used (*I do apologise* is an obvious one). Then ask the whole class which ways of apologising would be most suitable for each situation and practise saying it with appropriate stress and intonation.

3

1 Ask students to read the title and speculate how apologising can be a source of power.
2 Students read the text and match each paragraph with the correct topic. Give a maximum of one minute for this.

4 Ask students to read the stems of the seven questions and identify which paragraph they need to look in for the answer in each case. If the question contains a name such as Ben Renshaw they should look for the name in the text and underline it.
Then ask students to read the alternatives for each question and underline what they think are the important words. The questions here can be used to point out the kinds of similarities and differences they may find between alternatives. For example: for questions 1 and 2, ask students which two alternatives are comparatives, and for question 4 ask which of the alternatives talk about obligation and which talk about possibility.
Finally, ask students to read the text and choose the correct answers. Emphasise that they should go straight to the relevant part of the text in each case. Give about ten minutes maximum to complete the exercise before comparing answers in pairs.

5

1 Students divide the adjectives into two groups according to whether the meaning is positive or negative. If they are unsure, they should look again at the text to decide. Then students compare their lists in pairs. When going through the answers with the whole class, extend the exercise by asking students if they know the corresponding nouns, e.g. *arrogance*.
2 Students complete the sentences either individually or in pairs.

6 Students talk in pairs or groups about one or more of these situations. At the end, give them the opportunity to tell the class any interesting stories they heard.

ANSWERS

Ex. 3
2
1 C 2 A 3 B 4 F 5 E 6 D

Ex. 4
1 A 2 D 3 B 4 C 5 D 6 B 7 A

Ex. 5
1

self-righteous	N	honest	P	vulnerable	P
arrogant	N	heartfelt	P	committed	P
glib	N	guilty	N	single-minded	P
trivial	N	powerful	P	rewarding	P
proud	N	stubborn	N	insecure	N
fallible	N				

2
1 trivial 2 committed 3 guilty 4 proud
5 glib 6 stubborn

Grammar 2: gerunds and infinitives p.36

Aims:
- **to revise the use of gerunds and infinitives after certain verbs**
- **to focus on verbs followed by gerund or infinitive with a change of meaning**
- **to highlight the difference between present and perfect infinitive after such verbs**

1 Students look at the two questions, decide what they would do and then compare their choices with a partner. At the end, ask the class which of the three actions is the vindictive one.

2

1 Students read the article quickly, ignoring the gaps in order to answer the gist question.
2 Students work individually to put the verbs in either the gerund or the infinitive and then compare answers. Go through the answers with the whole class, building up two lists, verbs followed by gerund and verbs followed by infinitive, on the board. Ask students to suggest other verbs they know which could be added to the list.
3 Students look through the text to find the verb *allow*, which requires an object before the infinitive. Check that they understand that the object in this case is compulsory. Then ask them to find another verb in the text where a direct object before the infinitive is possible even though there is not one in this context (*prefer*).

3

1 Students work in pairs to consider each pair of sentences and answer the check questions. Then check the differences with the whole class.

2 Students again read the two sentences and answer the check question. Ask them to suggest some other sense verbs which could be followed by these structures. These could form other pairs to illustrate the difference between gerund and infinitive in this context (e.g. *I heard him call* versus *I heard him calling*).

4 This activity personalises the above grammar. Students complete the sentences and then compare and discuss them in pairs. Encourage them to ask follow-up questions about the sentences such as *Why do you avoid doing that?*

ANSWERS

Ex. 2

1

a) People who take revenge on others are acting naturally.

2

1 to sort 2 to get 3 to attack
4 to work out 5 to plan 6 planning
7 to get 8 to forgive
9 settling/to settle 10 finding

3

allows us to plan
3 and 9 prefer

Ex. 3

1

1 John, Peter 2 Jose, Carlos 3 Andrew
4 Jack, Jon

2

Susan

Speaking: language of possibility and speculation p.37

Aims:

- **to practise language used for speculating about relationships between people**
- **to highlight some useful language for talking about possibilities**

1 Students look at the three photos and discuss briefly in pairs what the relationship is in each case. Give about three minutes for this before comparing ideas as a whole-class activity.

2 Students listen to the recording and compare the ideas with their own. Pause the recording after the exchange about each photograph to ask students if they agree.

3 Students now listen again and complete the sentences. When checking the answers, ask one or two students to say each one with the appropriate stress and intonation.

4 Students now discuss two more photos and speculate on the relationships shown, using some of the above expressions and trying to improve on the language that they used in Exercise 1.

▶ Recording script p.93

ANSWERS

Ex. 3

1 it looks to me as if **2** guess is
3 get the impression **4** second thoughts
5 wouldn't be surprised
6 suppose it's just possible

Use of English: multiple-choice cloze (Part 1) p.38

Aim:

- **to complete an exam-style multiple-choice cloze (Paper 3, Part 1)**

1 Students discuss the two questions in pairs. This could lead into a general discussion about how important first impressions are and how they are created. It is often said that in a job interview, the first ten seconds are the most important.

2 Students read the title of the text and speculate briefly about the content. Then ask them to skim read the text, ignoring the gaps, to gain an overall idea of the content and see if their ideas are confirmed.

3 Students complete the multiple-choice exercise individually and then compare their answers in pairs. When checking the answers, draw attention to any useful collocations in the text such as *a great deal*, *scientific basis*, and *set out to prove*.

4 Students work in pairs or individually to complete the sentences. Emphasise that in many cases they will need to change the form of the word by adding a suffix.

5 This discussion is best done as a whole-class activity. It will probably highlight a number of points about body language, although if these were covered in the initial discussion in Exercise 1 above, you may prefer to keep it brief.

6 This can also be done as a whole-class activity. Question 2 is a good opportunity to point out that there can be cultural differences in this matter. For example, in the UK, not making eye contact is often seen as a sign that someone is not telling the truth; however, in some cultures, a lack of eye contact is a way of showing deference to the speaker.

ANSWERS

Ex. 3

1 B **2** D **3** A **4** D **5** B **6** B **7** A **8** D **9** B **10** C **11** A **12** B

Ex. 4

1 heartfelt **2** creation **3** contract **4** evolution **5** assess **6** conciliatory **7** communicating **8** artificial

Writing: information sheet (Part 2) p.39

Aims:

- **to complete an exam-style writing question (Paper 2, Part 2) requiring students to produce an information leaflet**

1 Students read the statements and decide which are true for an information leaflet.

2

1 Ask students to read the task carefully, underlining what they think are the most important phrases for successful completion of the task. Check their understanding of these by asking check questions such as *Who is the leaflet for?* and *Do you have to give positive or negative advice or both?*

2 In pairs or groups, students brainstorm possible ideas to include under these headings. After five to ten minutes, ask each group to report back on the ideas that they had and make lists for each heading on the board. This is an opportunity to weed out any ideas which may be irrelevant or misleading.

3 Students plan the leaflet in pairs, and decide on the title and headings. Encourage them to use different or additional headings from the ones given rather than simply copying them.

4

1 Students read the example answer and discuss the questions together. Then go over the questions with the whole class, pointing out any useful pieces of language that are used to introduce the advice, such as *However – a word of warning* or *There is nothing worse than …*

2 Students read the leaflet again and identify spelling mistakes. They compare their corrections in pairs before checking as a whole-class activity.

3 Again students re-read the leaflet and identify the two grammar mistakes. Point out that when they are checking their work for errors, it is a good idea to read it two or three times and look for a different kind of error each time: once for spelling errors, once for tense errors and so on.

5

1, 2 This can be done in class or for homework. If it is given for homework, students can swap and read each other's leaflets in the following lesson. Ask students to read their partner's leaflets at least twice, firstly looking at the overall layout and organisation and then more closely to check the grammar and spelling. You could practise the piecemeal editing technique suggested above by asking them to proofread once for spelling and once for grammar or verb forms.

Teaching tips and ideas

Students evaluating each other's work, both to check for errors and also for feedback on the content, is something which can be introduced on a regular basis. It improves students' ability to monitor their own work and provides them with a number of example answers to any writing task. One simple technique is to ask them to tell their partner one thing that they thought was particularly good in his/her answer and one phrase or sentence that seemed particularly well expressed. These can then be shared with the whole class at the end of the activity.

ANSWERS

Ex. 1

1 F **2** T **3** T **4** F **5** T **6** F

Ex. 4

1

1 a)
2 Yes – it establishes the purpose of the leaflet
3 Yes
4 Yes – talks directly to the reader
5 They make the *dos* and *don'ts* stand out, not to overuse them

2

~~Wat~~ – What ~~creat~~ – create
~~uncomforetable~~ – uncomfortable
~~acheive~~ – achieve
~~friendlyness~~ – friendliness
~~monosylabic~~ – monosyllabic
~~advise~~ advice
~~maintan~~ – maintain

3

you ~~couldn't~~ *shouldn't* dress down too much
Remember you want ~~conveying~~ *to convey* an impression …

UNIT 3 Review p.41

ANSWERS

Ex. 1

1 embarrassing **2** uncontrolled **3** ridiculous **4** increasingly **5** destructive **6** Consequently **7** Intolerance **8** intake **9** aggression **10** advisable

Ex. 2

1 should **2** must **3** have **4** have, can **5** can **6** will

Ex. 3

1 Our brains allow us ~~planning~~ *to plan* our lives well, which animals can't do.
2 I always try to ~~working~~ *work out* the best solution to problems by talking them over with friends.
3 [correct]
4 When people take chances, they can risk ~~to find~~ *finding* themselves in difficult situations.
5 [correct]
6 I really regret not ~~to have~~ *having* studied harder when I was at school.

UNIT 4 Pushing the boundaries

Vocabulary 1 p.42

Aim:

- **to complete an exam-style open cloze**

1 Begin by writing the word *science* on the board and asking students to name different branches such as biology, astronomy and so on. Then use one or more of the questions here to conduct a brief class discussion.

2 Students briefly speculate on the content of the text and then skim read it to confirm their predictions.

3

1. Students complete the exercise individually or in pairs before checking as a whole-class activity.
2. Students find the words and phrases from the text. Ask them if they can suggest some other common collocations for the phrasal verb *break down*.

4 This is best done as a whole-class discussion. Ask students for examples of the good or the bad effects that science has produced now and in history.

ANSWERS

Ex. 2
Its unpredictability and the fact that many discoveries are made by chance

Ex. 3
1
1 by/with 2 other 3 no/little 4 all 5 up 6 why 7 is 8 make 9 it 10 what 11 which 12 into 13 former 14 most 15 never
2
1 thrown up 2 make sense of 3 break down 4 for the most part

Speaking: Parts 3 and 4 p.43

Aim:

- **to focus on strategy for answering exam-style speaking tasks (Paper 5, Parts 3 and 4)**

1

1. Students listen to the recording and summarise the instructions. Point out that there are two elements involved, having a discussion and making a decision.
2. Students listen and say why the two candidates are not answering the task.
3. Students match the phrases individually or as a whole-class activity.
4. Students listen to the conversation and identify which expressions are used. Point out that these two students are carrying out the task correctly because they are giving and explaining opinions.
5. Students now complete the speaking task in pairs. You could ask them to make sure that they use at least two of the expressions in task 3.

2 Students read questions 1 to 6 individually and take a few seconds to think about them. Then they listen to the recording and talk in pairs about how the candidates' opinions differ. They then listen again to pick out the phrases from Exercise 1.3 and note down any additional phrases. Finally, they discuss the other questions, giving about two to three minutes for each one. You might like to ask one pair to discuss question 2 first in front of the class so that the class can comment. It is also worth pointing out the importance of examples in justifying opinions, like candidate B's example of medicine.

3 Students discuss the questions using phrases they heard in the recording.

ANSWERS

Ex. 1
3
Clarifying: So what you mean by that is ...; So you're saying that ...;
Asking: How do you feel about ...; Do you feel the same as ...; What do you think about ...
Explaining: What I mean is ...; I feel that ...; I'm trying to say that ...; It seems to me that ...

Ex. 2
3
Phrases from Exercise 1.3: I feel that ...; it seems to me that ...
Phrases that add information: And what's more ...; Another thing I think about ...; On top of that ...

▶ Recording script p.93

Grammar 1: conditionals (overview) p.44

Aim:

- **to review the structures used in conditional sentences and provide spoken practice**

1

1 At this level, students should already be familiar with the basic three conditional types. They work individually to complete the sentence transformations and then compare in pairs.
2 Elicit the rules from the whole class, which students complete for reference. Ask the class for examples of each rule from the sentences in 1.

2 Students correct the mistakes either in pairs or individually. When going through the answers, ask students to pronounce the contracted forms such as *I'd've done better*.

3

1 Students briefly discuss if they think the possible changes will happen (e.g. *Do you think that cosmetic surgery will become cheaper?*) and then decide on a first conditional sentence for each one. Conduct a class feedback by asking each pair to provide one first conditional sentence.
2 Begin this activity with a quick brainstorm. Write the three given areas on the board and ask students to suggest 'unlikely' changes, using their imagination (e.g. *If they invented a car which ran on water, the pollution problem would be solved.*). Then students work in pairs to write conditional sentences for the three topics.
3 Students work individually to write at least three third conditional sentences, beginning *If … had not been invented …* and then read their sentences to each other. Again, encourage contracted forms when speaking.

4 Give students about five minutes to complete the sentences individually, before reading them to each other in pairs or groups. Encourage them to ask further follow-up questions.

ANSWERS

Ex. 1
1 don't do, you will 2 had known
3 would use, had 4 (automatically) adds milk, press this button 5 you come, stand up
6 you touch, might

2
1 could, might, may … will, would
2 present
3 *if* + past perfect + *would*
4 *if* + past + *would*
5 second
6 *if* + present + *will*

Ex. 2
1 If you really want to keep up with scientific developments you **will** have to read more!
2 If there were more information about science on television, there's a chance that young people **might** get interested in it.
3 There is a great offer on sci-fi books on the Internet – if you buy two you **get** one free.
4 I would have done better at science when I was at school if I **had** worked harder.
5 If I promised to take care of it, **would** you lend me your video mp3 player?
6 I wouldn't take that job **if** I were you!
7 If he **had gone** to the party, he might have seen her there.
8 If you had taken up her offer of a lift, you might **have** got home sooner.

Exam focus

Paper 4 Listening: multiple choice (Part 1) p.45

Aim:

- **to complete an exam-style listening task (Paper 4, Part 1)**

Go over the exam information and exam procedure with the students. Then ask students to read the two questions for extract one. Remind them that the actual words on the recording are likely to be different from those in the questions and ask them to suggest alternative ways of expressing some of the ideas here such as *disappointment* and *salary*. Then play the recording twice for the first extract and ask students to compare their answers. Check them together while they are still fresh in the students' minds. You might like to play the recording a third time for this and ask students to identify points at which they can eliminate the incorrect answers (stage 3 of the procedure).
Follow the same procedure for extracts 2 and 3. For question 5, which focuses on the function of what the speaker is saying, ask the students to suggest language that might be used for apologising, blaming or explaining.
With a strong group, you could play the extracts straight through and check all the answers at the end. This makes the task more similar to what the students will do in the exam, but there is less chance to check that students are using the suggested procedure.

ANSWERS
1 C **2** A **3** C **4** B **5** C **6** B

▶ Recording script p.94

Reading: gapped text (Part 2) p.46

Aim:
- **to complete an exam-style gapped text reading**

Exam information

In Paper 1, Part 2, candidates read a text from which six paragraphs have been removed and are required to re-insert the paragraphs in the correct place. This tests their understanding of the overall text structure. For this task, students will need to develop their awareness of cohesive devices such as link words, referencing devices and synonyms.

1 Use one or both of the questions to introduce students to the topic. You could also ask students if they know any sayings which express an optimistic or pessimistic point of view. Examples in English might be *Everything happens for the best* versus *If something can go wrong, it will.*

2

1. Students skim the first paragraph and predict what the writer will say about Murphy's Law.
2. Students first read the whole text, ignoring the gaps. Then they read the missing paragraphs A–G. Now ask them to look again at the base text and underline any link words at the beginnings of the paragraphs such as *despite*. These will link back to something in the content of the missing paragraphs. Ask if they can see any other words or phrases which they think must link back in this way (e.g. *such examples* in the paragraph after gap 4). Now ask students to work individually for about ten minutes to put each missing paragraph in the correct gap. They should read through the base text, stopping at each gap in turn to decide which paragraph is most suitable. At the end, they should re-read the whole text through. Finally they compare their version with a partner. (A detailed suggested procedure for this type of exercise is given in Unit 8.)

3 Students first do the exercise without looking at the text. Then they refer to the context in the text.

4 Students discuss the equivalent of Murphy's Law in their own language, and go on to talk about their personal opinion.

ANSWERS

Ex. 2
2
1 C 2 E 3 D 4 G 5 B 6 A

Ex. 3
1 g 2 f 3 e 4 c 5 h 6 b 7 a 8 i 9 d

Listening: sentence completion (Part 2) p.47

Aim:
- **to complete an exam-style listening task**

1 Ask students to read the introduction and the title. Then they read the gapped sentences. Check their understanding of what kind of information is required for each gap by asking questions (e.g. *Which answer is a job?*) Then ask if they can predict any likely answers. Check students understand that they should not write more than one or two words for each gap. Then play the recording and students note the answers. They compare in pairs before listening again to check.

2 This may be just a brief discussion but in some groups it could lead to a longer discussion on the possibilities of genetic engineering and its problems.

Exam information

In CAE Paper 4, Part 2, students listen to a talk and complete sentences which summarise the content. The gaps require them to understand specific information or occasionally stated opinion. Before they hear the recording, they should look through the sentences, thinking about what kind of information is needed in each gap and try to predict likely answers by considering the collocations and context.

ANSWERS

Ex. 1
1 biology 2 popular science
3 (thorough) research 4 schoolteacher
5 feathers 6 cover 7 genetic engineering
8 shellfish

▶ Recording script p.95

Vocabulary 2: word formation p.48

Aim:
- **to complete an exam-style word-formation task**

1 Students read quickly about the four discoveries and try to identify them. If they do not know the name of the drug in text C, ask them which disease they think it cures.

2 Students complete the exercise either individually or in pairs. You may wish to elicit word diagrams for some of the words here, especially *botany, miracle, diagnosis* and *photograph*.

3 These questions are best answered as a whole-class activity.

ANSWERS

Ex. 1
A penicillin B anaesthetic C quinine D x-ray

Ex. 2
1
1 breakthroughs 2 previously 3 resistance
4 growth 5 Botanists 6 incredible
7 accidentally 8 miraculously 9 diagnostic
10 photographic
2
1 breakthrough 2 previously
3 diagnostic, photographic
4 accidentally, miraculously
5 resistance, growth 6 botanists 7 incredible

Grammar 2: conditionals (advanced) p.48

Aim:
- **to introduce more advanced conditional structures, including inversions, *happen to* and alternatives to *if***

1
1 Students may have encountered some but probably not all of the structures here before. They work individually to tick the options that they feel are possible and then compare in pairs or groups. Then go though the answers with the whole class.
2 If students already knew most of the structures in Exercise 1.1, they could do this exercise in pairs. Otherwise, it is best done as a whole-class activity.
3 This is again best done as a whole-class activity.

Watch Out! *in case* and *if*
This note aims to clarify the difference between *in case* and *if*.

2 Students work in pairs to insert the missing words. If they have difficulty, ask them to look again at the examples in Exercise 1.1

3 This can be done individually or in pairs.

4 Students talk in pairs or small groups to discuss the two dilemmas and report back to the class about what they would do.

ANSWERS

Ex. 1
1
1 a) ✓ b) ✓ 2 a) ✓ b) ✓ 3 a) ✓ b) ✓ d) ✓
4 a) ✓ b) ✓ c) ✓ 5 a) ✓ b) ✓ 6 b) ✓ c) ✓
2
1 3a, 3d
2 4a
3 2b
4 2a
5 1a, 4c
6 5a
3
b – these words may stress the hypothetical nature of the conditional clause, but there is no difference in politeness or formality.

Ex. 2
1 **Had** we spent …
2 If you **will** just …
3 If I were **to** say …
4 If you happen **to** see …
5 … **would** that be a problem?

Ex. 3
1 … **unless I am** sure it's safe.
2 … **had you** taken my advice.
3 **Were someone to find** a cure …
4 … if you **happen to find** them …
5 … **provided that** the weather improves.
6 **As long as** she works hard …

▶ Photocopiable activity 4A *Matching conditionals* p.160

Vocabulary 3: collocations, fixed phrases and idioms p.50

Aim:
- **to review collocations, fixed phrases and idioms and to highlight some common examples**

1 If students are not familiar with the concept of collocation, use the example *catch a cold* and ask which other nouns commonly go with the verb *catch* (e.g. *a fish, fire, a glimpse*). Then students choose the correct verbs for questions 1 to 8.

2 This exercise focuses on adverb–adjective collocations in the same way.

Teaching tips and ideas

Recording collocations should form an ongoing part of students' vocabulary notes. Again, this can be done using the word diagram format with a common verb, noun or adjective as the base word and common collocations arranged around it. This is more memorable than recording them in a list.

3 Go over the explanation of the concept of fixed phrases and then ask students to match the sentence halves. Ask them to suggest other fixed phrases using the same first nouns (e.g. *pack of cards, point of law*).

4

1 At this level, students will probably be familiar with the idea of idiomatic speech, and so you might elicit a definition from them before going over the one provided here. Then ask students to read the idioms a to e and discuss in pairs what they think they mean. Then they read sentences 1 to 5 to answer the questions. You might like to allow them to use a dictionary to check their answers before checking with the whole class. You may want to introduce students to some additional idioms (e.g. *let the cat out of the bag, keep a low profile, play your cards close to your chest*), which they can then discuss with a partner.

ANSWERS

Ex. 1
1 hold their breath 2 twist their ankle
3 pull a muscle 4 make a good living
5 hack into computers
6 bookmark your favourite web pages
7 prioritise your work 8 conduct experiments

Ex. 2
1 utterly 2 bitterly 3 hugely
4 deeply 5 enormously 6 completely

Ex. 3
1 b 2 c 3 a 4 e 5 d

Ex. 4
2 c 3 b 4 a 5 e

▶ Photocopiable activity 4B *Idiom call my bluff* p.161

Exam focus
Paper 3 Use of English: gapped sentences (Part 4) p.51

Aim:
- **to introduce students to an exam-style gapped sentences task**

Go over the exam information and procedure with students. Students can either complete the task individually, or you may choose to do the first two questions together as a class. At the end, ask students to suggest how they can help themselves with this type of task (noting contexts and a variety of common collocations for words in their vocabulary notes).

ANSWERS
1 short 2 blocked 3 deal 4 turn 5 process

Writing: article (Part 2) p.52

Aim:
- **to give practice in writing an article in response to an exam-style writing task**

1

1 Students discuss together which of the statements are true for articles and then check them against the article on page 46.
2 Students discuss in pairs or groups to choose two of the pieces of advice here and then report their choices to the class. Paragraphs are of course necessary although they would probably not count as a means of making the article interesting.

2

1 Give students a few moments to read the task carefully and then check their answers to the three questions.
2 If students are already practised at writing articles, you could ask them to write a short introductory paragraph of their own to compare with the two examples. Otherwise, students read the two introductions and tell you which is more appropriate.

3

1 You can either give students a quiet two or three minutes to think of some ideas or allow them to brainstorm in groups. You will need to monitor the ideas that they come up with to check that they are relevant and not just rewordings of the same idea.
2 Students now write up the their points into three paragraphs.

3 Students now write the conclusion. Again, you will need to check that the conclusion is not overlong and does not include anything which is completely different to the previous three paragraphs.

4 Students can either proofread their own articles or in a supportive class, they could check each other's. Remind them of the piecemeal editing technique of looking for one type of error at a time.

ANSWERS

Ex. 1
Not true: should have bullet points and headings

Ex. 2
1
1 support or disagree, give reasons
2 young people
3 informal, direct
2
1 a
2 b – it's more direct, more informal and goes straight to the point.

UNIT 4 Review p.53

ANSWERS

Ex. 1
1 in **2** from **3** towards/for **4** like **5** the
6 to **7** which **8** of **9** No/Little **10** part
11 up **12** so **13** As **14** only/just **15** all

Ex. 2
1 c **2** b **3** b

Ex. 3
1 keep in touch (informal)
2 taking a break (informal)
3 spots (informal)
4 set up (informal)
5 signed up (informal)
6 growing involvement (formal)

Ex. 4
1
technological – adjective; other forms – technology
warn – verb; other forms – warning
revolution – noun; other forms – revolt
expertise – noun; other forms – expert
evidence – noun; other forms – evident
genuine – adjective; no other forms
2
1 warning **2** technical **3** expert
4 revolutionary **5** genuinely

UNIT 5 Thrills and skills

Listening 1: multiple matching (Part 4) p.54

Aims:

- **to give practice in listening for opinion and attitude**
- **to give spoken practice in explaining rules and processes within the context of sports**

1 Students note their answers to the questions and then compare them in pairs.

2 Students listen to the recording and match the topics. When going over the answers, stop after each extract and ask students to summarise the speaker's opinion.

3

1. Students now listen and make notes. Check the answers as a whole-class activity, highlighting any differences between their ideas and the speakers'.
2. Students listen again and note the names of the unusual sports mentioned.

4

1. This can be done in either pairs or groups depending on the size of the class. Encourage them to think of precise rules for the new sport, which need not be exactly the same as the rules in the original two.
2. Students could either tell the other groups about the sport informally, or you could provide them with marker pens and paper or OHTs and ask each group to prepare a short presentation before voting.

ANSWERS

2.1

1 B 2 A 3 D 4 C 5 F

3.2

bungee running, bouncy boxing, boxercise

▶ Recording script p.96

Grammar 1: intensifiers/modifiers p.55

Aim:

- **to test and extend students' knowledge of the use of modifiers and to provide controlled practice**

1

1. Ask students to look at the sentences and check that they know the meaning of the term *modifier*. Then they suggest other modifiers that could be used.
2. Students work either individually or in pairs to complete the exercise. When checking the answers, point out that some alternatives are incorrect because modifiers such as *totally* can only be used with non-gradable adjectives. Others are a matter of collocation. Students should be encouraged to record common modifier–adjective collocations in their vocabulary notes (e.g. from this exercise *absolutely clear*, *absolutely amazing*, *absolutely overwhelming* and *completely honest*).

2

1, 2, 3 Students now use their answers to Exercise 1.2 to help them sort the adjectives into two groups and then follow the same procedure with the modifiers. Point out that *really* can be used with some gradable and non-gradable adjectives, as can *quite*, with different meanings. *Quite* means 'fairly' with gradable adjectives and 'totally' with non-gradable ones. Emphasise again the importance of recording collocations here, as *quite* is not used with all non-gradables.

3 Students work individually to find the mistakes and then compare in pairs.

4 Allow students a short time to think and then ask them to talk in pairs about one of these times. To shorten the exercise, you could ask them to choose just three or four collocations.

ANSWERS

Ex. 1

1

To be perfectly honest

Professional sport is very conservative at heart

2

1 absolutely 2 fairly 3 very/really

4 really/absolutely 5 quite/utterly 6 extremely

7 very 8 absolutely/totally 9 completely

10 extremely/terribly

Ex. 2
1
1 G 2 G 3 G 4 U 5 U 6 G 7 G 8 U
9 G 10 G
2
Gradable: very, terribly, rather, extremely, fairly
Ungradable: absolutely, completely, totally
3 **Really**: can be used with both

Ex. 3
1 I found the whole situation **rather** embarrassing.
4 She felt **extremely** nervous before going on stage.
6 It should be very clear that the situation is **very** difficult.
8 I find the plan **completely** acceptable.

Exam focus

Paper 1 Reading: multiple matching (Part 4) p.56

Aim:

- **to introduce and give practice in reading techniques for answering CAE Paper 1, Part 4**

Exam information

In Paper 1, Part 4, there is a text, usually divided with subheadings, and 15 questions. Students are required to match each question with the correct section of the text.

Ask students to look at the task and suggest what reading techniques they should use. If necessary, prompt them with questions such as asking them whether they should read the text or questions first. Then go over the exam information and suggested procedure. For stage 5, emphasise that more than one text may say something similar to the question, so that at this final stage, students have to think carefully about the meaning before they make their choice.

1 Students now complete the reading task individually.

2 Students compare their answers and where they found them. They discuss any differences and make a final choice together. When going through the answers, point out that the language in the questions will often be different from the language in the texts so that students need to be prepared to spot synonyms, such as *crashes* for *accidents*. It is also worth pointing out that the language in the texts will often be specific whereas the questions express ideas in a more general way such as *topples out* for *has the occasional mishap* in question 5.

3 Students complete the matching exercise individually.

4 This can be done as a whole-class discussion, or you could ask students to brainstorm disadvantages in pairs.

Teaching tips and ideas

As students need to get used to reading under time pressure, you may like to set a time limit of around 20 minutes. If students find this difficult, start with a slightly longer time and gradually reduce it in subsequent lessons. If available, a stop watch is useful for this.

ANSWERS

Ex. 1
1 D 2 A 3 C 4 D 5 B 6 C 7 D 8 A
9 D 10 B 11 B 12 A 13 B 14 C 15 A

Ex. 3
1 i) 2 c) 3 g) 4 h) 5 d) 6 e) 7 a)
8 f) 9 b)

Vocabulary 1: word formation (prefixes) p.58

Aim:

- **to revise and extend students' knowledge of negative prefixes and provide controlled practice**

1 Students complete this initial exercise in pairs. Most students will probably be able to think of two words with these prefixes without using a dictionary but to extend the exercise, you could ask them to find one additional word for each prefix in the dictionary.

2 Students again work in pairs, with one person reading the given sentence and the other supplying the contradiction. With a strong group, this exercise provides an opportunity to introduce the concept of shifting word stress. In a contradicting sentence such as in the example here, in spoken English, the stress on the contradicting word will change from its normal position to the prefix. Illustrate this with two short exchanges, as follows:

A I think the new stadium has very hard seats.
B Yes they're not very ***com****fortable.*

A I think the new stadium has very comfortable seats.
B Really? I think they're very ***un****comfortable.*

Students should then be asked to stress the prefix in the contradicting exercise.

3 Ask students to work individually to match the meanings of the prefixes to the correct sentences and then compare in pairs.

4 Students now work in pairs to decide on the correct prefixes and write example sentences.

ANSWERS

Ex. 1 sample answers
unfortunately/unhappy, dishonest/disconnected, immobile/impossible, illegible/illegal, irrational/irresistible

Ex. 2
1 I think it's irrelevant.
2 I think they were illogical.
3 I think he seemed immature.
4 I think they are more disobedient.
5 I found it unbelievable.

Ex. 3
1 b) 2 h) 3 e) 4 g) 5 c) 6 j) 7 a)
8 d) 9 i) 10 f)

Ex. 4
postgraduate/undergraduate reinstate anticlockwise misunderstand understatement/overstatement underactive/overactive/reactive

▶ Photocopiable activity 5 *Prefixes* pp.162 and 163

Use of English: word formation (Part 3) p.58

Aim:
- **to provide practice in completing an exam-style word-formation exercise**

1
1 The word *counterfactual* is itself an example of a word whose meaning might be inferred from the prefix. Prompt students to guess the meaning by giving other examples of words with the prefix *counter* or *contra* such as *counterargument* or *contradict.*
2 Students now read to find out the meaning of *counterfactual* as explained in the text. Discuss whether they think it is a true description of how people think.

2 Students now complete the word-building exercise either individually or in pairs. When checking the answers, emphasise that they must be exactly correct, with correct spellings and the plural 's' on numbers 3 and 8.

3
1 If students find it difficult to relate the two types of thinking to specific people, you could just ask them to provide further examples of situations in which people are likely to think in a conterfactual manner. Receiving exam grades or passing or failing exams are an example that students could probably relate to.

ANSWERS
1 closeness 2 satisfaction 3 medallists
4 frustrating 5 inactivity 6 unwise
7 powerful 8 adjustments 9 uncomfortable
10 rewrite

Speaking: agreeing and adding information (Parts 3 and 4) p.59

Aims:
- **to practise discussing possibilities and reaching a decision**
- **to introduce phrases for expressing partial disagreement or introducing additional ideas**

1 Students look at the task and suggest some ideas about why the different possibilities would be effective.

2
1 Students now listen to a discussion on this topic. For the first listening they should concentrate on understanding the content and note down the most important points. At the end of the listening they compare in pairs.
2 Students listen to the discussion a second time to focus on the phrases for introducing additional points and for expressing disagreement. You may play the discussion all the way through first to see how many students can pick out the expressions, and then repeat it, pausing the recording after each relevant phrase. List the phrases on the board as you go through, then ask students to suggest any other phrases which could be used.

▶ Recording script p.96

3 Students now complete the speaking task in pairs. Set a time limit of about five minutes for them to reach a final decision.

4 This discussion activity is best done in pairs. Ask one student in each pair to note down at least two reasons for their opinion (or two differences in the case of question 3) so that they can tell the class their ideas at the end. Weaker groups could be asked to think individually about the questions and make some notes before discussing with a partner.

ANSWERS

2.2
Make an additional point:
On top of that …
What's more …
Not only that, but …

Indicate partial agreement:
Having said that …
That's all very well, but …

Agree with a point:
I take that point on board.
You could be right.

Grammar 2: intensifying comparative forms p.60

Aim:
- **to focus on comparative structures and to provide controlled practice in using modifiers and intensifiers**

1
1, 2 Students complete the two exercises individually or in pairs.
3 They choose formal and informal expressions and again compare the ones they chose. When going through the answers, encourage them to record any useful collocations and phrases in their vocabulary notes, e.g. *considerably better, nothing like as much.*

2 This speaking task could be made more structured by giving students adjective prompts on the board, e.g. *exciting, fun, relaxing.*

ANSWERS

Ex. 1
1
1 great deal 2 much more 3 considerably 4 by far the
2
1 not nearly as 2 a lot more 3 rather 4 loads 5 half as many 6 a bit 7 nothing like as many 8 more and more 9 slightly
3 formal: considerably better, a great deal, by far
informal: a lot more, loads, a bit

Vocabulary 2: sports idioms p.61

Aim:
- **to introduce students to some idiomatic expressions involving sports vocabulary**

1 Ask students to work individually to read the statements and match each one with the correct person. They then compare answers in pairs. If they do not know the idioms, encourage them to guess.

2
1 Again, students work individually and then compare in pairs. The idioms here are rather more difficult to guess, so you may want to allow students to check in a good dictionary or an idiom dictionary.
2 When going through the answers to 2.1, ask students to identify which sport they think the idiom comes from.
3 This can be done as a whole-class activity.

3 Students now write their own example sentences for the given idioms. They will probably need to check the meanings in a dictionary first. If they do not have access to an idiom dictionary, you could provide a list of definitions on the board or OHP, which they then match to the correct idiom. At the writing stage, you will need to monitor the students' sentences to make sure that the idioms are correctly used.

4
1 This exercise reinforces the meanings of the idioms in Exercise 2. You might start by describing an experience of your own and asking which of the idioms could be used to describe it. Then give students one or two minutes to think of examples of their own before talking in pairs.
2, 3 This exercise also gives students the opportunity to discuss the careers of famous sportspeople. The opportunities for discussion will probably be greater in a multinational class where students can tell each other about well-known sportspeople in their own countries. In a monolingual class, you could talk more generally about what drives sportspeople and what kind of careers they have.

ANSWERS

Ex. 1
a) 3 b) 6 c) 4 d) 5 e) 2 f) 1

Ex. 2
1 b) board game, e.g. chess
2 e) betting – horse racing
3 a) archery/darts
4 i) tennis
5 c) swimming/diving
6 d) car racing/Formula One
7 f) athletics/racing
8 g) football
9 h) football/baseball/any team ball game

3
1 goes off the deep end
2 the ball is in your court
3 above board
4 succeeded against all the odds

Use of English: open cloze (Part 2) p.62

Aim:

- **to practise techniques for completing an open-cloze exercise for CAE Paper 3**

1 If students are not familiar with this topic, you may need to preteach the words *glide* and *glider*. Ask the questions here to the whole class to introduce them to the topic, encouraging them to guess if they do not know.

2

1 Students skim the text quickly to answer the global multiple-choice question. Give a time limit of about one minute for this.

2 Students work in pairs or individually to complete the exercise. When going through the answers, point out that while some answers depend on fixed phrases or collocations such as 6 (*take advantage*), others depend on their ability to link the ideas across sentences and paragraphs within the text. For example, question 12 depends on their ability to relate the word *flexibility* to the situation described in the previous sentence and question 10 relies on the previous reference to *engineless aircraft* in paragraph one.

3, 4 These questions may be asked to the whole class to round off the activity.

ANSWERS

Ex. 2
1 b)
2
1 whose 2 before 3 up 4 as 5 what
6 take 7 which/that 8 one 9 at 10 no
11 their 12 this/such 13 off 14 many 15 in

Listening 2: multiple choice (Part 3) p.63

Aim:

- **to complete an exam-style multiple-choice listening task**

1 Write *indoor climbing* on the board and ask students to speculate on what it could involve, using the three questions.

2

1, 2 Ask students as a class to find the words and then check any other expressions such as *time on your hands* which they may not be familiar with.

3 Give students another minute to re-read the questions and then play the recording. They check answers in pairs before listening a second time.

4 Briefly ask for students' opinions on the sport described in the listening.

▶ Recording script p.96

ANSWERS

Ex. 2
1
a) basic training = grounding
b) beginners = novices
c) a social grouping = sub-culture
d) not very willing = reluctant
e) support one thing at the expense of another = take sides

Ex. 3
1 C 2 B 3 D 4 A 5 C 6 D

Writing: a reference (Part 2) p. 64

Aim:

- **to give practice in writing a reference in response to an exam-style writing task**

1 Ask students to read the task carefully and check they understand the nature of the job and who will read the reference.

2 Students read the task and decide which points are suitable, with reference to their previous discussion.

3 Students read the example answer on p.191 and identify the two irrelevant sentences. Then they work individually or in pairs to do the vocabulary exercise.

4 Students work individually to complete the plan. Go through the answers, checking that students understand how the different tenses correspond to different typical features of a reference.

5 This could be set for homework, or students could write the references in class and then read each other's and comment.

ANSWERS

Ex. 2

2

– what you and your friend have done together
– reasons why you like your friend
– informal or colourful language

Ex. 3

1

Irrelevant sentences:
She is a good friend of mine and we have been at school together for most of that time.
She enjoys reading and is particularly keen on science fiction, which she reads all the time.

2

1 courteous 2 accustomed to 3 proficient at
4 In addition 5 In the past 6 at a high level
7 consequently 8 therefore 9 invest
10 I can highly recommend her for the position

Ex. 4

Para. 1: present
Para. 2: + her character; present perfect/present
Para. 3: skills and qualifications; present/past/hypothesis
Para. 4: Future plans
Para. 5: recommendation; hypothesis

UNITS 1–5 Progress test p.65

The progress test section follows the format of CAE Paper 3 (Use of English). The exercises could be set for homework or done as further practice in class.

ANSWERS

Ex. 1

1 D **2** B **3** C **4** A **5** B **6** B **7** C **8** A
9 C **10** B **11** D **12** A

Ex. 2

13 where **14** of **15** had/needed
16 lack/want **17** at **18** If
19 more/greater/further
20 While/whilst/(al)though
21 much **22** not **23** what **24** all **25** ought
26 like **27** well

Ex. 3

28 handsets **29** typically **30** upgrading
31 consumer **32** emotional **33** unwilling
34 donation **35** reconditioned **36** affordable
37 reliable

Ex. 4

38 saving **39** apply **40** flat **41** study **42** rare

Ex. 5

43 gave a beautiful performance
44 led to an/the increase in/led to the increasing
45 who has organised
46 it that makes
47 I would have been able
48 cross the finishing line
49 no means unusual
50 had not gone out so

UNIT 6 Family ties

Reading 1 p.68

Aim:

- **to give practice in reading techniques (skimming and scanning, followed by careful reading) for Paper 1, Part 4**

1 Students work in pairs to make a list and then classify the issues. Teenage students or students who are the parents of teenage children may be able to give good examples but the topic needs careful handling as personal issues may emerge.

2 Students skim the text to answer question 1, within a time limit of one minute. Before they attempt 2.2, remind them of the procedure for answering this type of exercise, (read the questions first carefully, then scan the text for the answers one by one). You may need to teach the verb *counter.*
If appropriate to the class, the topic could be personalised with students discussing incidents from their own life in pairs (e.g. talk about a time when they won an argument as a teenager, were allowed to so something for the first time, or not allowed to do something that their friends were).

ANSWERS
1 D **2** A **3** C **4** B **5** A **6** D **7** D **8** C
9 B **10** B

Listening 1 p.69

Aims:

- **to give practice in listening for attitude and opinion**
- **to give practice in making suggestions and agreeing and disagreeing in spoken English**

1 Give students time to read the questions carefully for the first extract and then play the recording. They compare answers in pairs before listening again. Follow the same procedure for extract two.

▶ Recording script p.97

2

1. Do the first part about rules for parents together and write the rules that students suggest on the board (e.g. *do not argue in front of the children*). Students then work in groups to decide on rules for siblings and grandparents.
2. Different groups now join up to compare rules and decide on the best four. Again, you may need to consider students' ages and backgrounds when forming the groups. The topic could be consolidated by asking students to write a report on their discussion (e.g how they agreed on the rules, what rules they agreed on and how they think the rules will work). This could form the basis of a presentation made by each group to the class. Give each group two or three OHTs or sheets of paper which they use to make visuals to illustrate the rules. Ask the class to suggest some ways to ensure that the rules were kept (e.g. monthly family meetings). Alternatively, it could be given as a report-writing exercise for homework.

ANSWERS
Ex. 1
1 C **2** A **3** C **4** B

Grammar 1: hypothetical meaning – *wish* p.70

Aim:

- **to revise and extend students' knowledge of structures used after *I wish* and the use of unreal past after *it's time, would rather* and second conditional structures**

1 Students do this exercise individually or in pairs. Go through answers, checking that the class understands the use of the past simple after *would rather* and that *it's time* does not refer to the past. At the end, ask which sentences hypothesise about an imaginary present or future (1, 2, 3, 7, 8) and which one refers to an imaginary past (6). Finally, students work in pairs to match each sentence to a rule.

2

1. Use the text first to practise skim reading by asking students to look through the text quickly to find the family relationship. You could also ask what difficulties each person mentions. Then students choose the correct verb forms for their text.
2. Students summarise the content of the texts to each other and check each other's verb forms.

3 This exercise gives some freer practice of this grammar area.

4 This can be done as a whole-class activity. In a multilingual class, the activity could lead to cultural comparisons on the size and types of families in different countries.

ANSWERS

Ex. 1

1

1 I wish my dad ~~couldn't~~ wouldn't always be so cross with me.
2 I wish I ~~have~~ had an older brother.
3 (correct)
4 Would you rather I ~~will call~~ called back later?
5 It's definitely time the children ~~go~~ went to bed.
6 (correct)
7 I wish I ~~would~~ could stop eating chocolate!
8 Suppose nobody ~~will come~~ came to the party – I'd be really disappointed!
9 I really wish I ~~can~~ could play the guitar!

2

1 b) 2 a) 3 d) 4 f) 5 e) 6 h) 7 c) 8 g) 9 i)

Ex. 2

1

Student A:

1 have 2 wouldn't worry 3 coped 4 didn't complain 5 hadn't moved in

Student B:

1 would slow down 2 didn't 3 would turn 4 had turned him away 5 waste 6 tried 7 could tell

Exam focus

Paper 5 Speaking: collaborative task/ discussion (Parts 3 and 4) p.70

Aim:

- **to give practice in carrying out Parts 3 and 4 of the CAE speaking exam**

1

1 Go over the exam information and procedure. Then students listen to the interlocutor's instructions and identify the task.

2 As well as deciding which candidate does better, students should also give examples of the ways in which they initiate discussion and any good language used.

▶ Recording script p.98

2

1 Students complete the speaking tasks in groups of three with one student as the interlocutor. If your group is not divisible by three, have some groups of four with one student as the assessor.

2 Conduct a brief feedback session after students have done the speaking tasks by asking how candidates and interlocutors feel they did.

3 When repeating the task, it is best to reform the groups completely, making sure that all the previous interlocutors are now candidates.

Vocabulary 1: word + preposition(s) p.71

Aim:

- **to revise and extend students' knowledge of prepositions used after certain verbs**

1 This is best done in pairs so that students pool their knowledge. They could also use dictionaries to check the prepositions for any verbs that they do not know.

2

1 Students skim the text quickly and answer this gist question.

2 Students work in pairs to insert the prepositions. Check the answers and build up the list of verbs and prepositions on the board.

3 Use this as a general discussion with the whole class.

4 This exercise introduces the idea of different prepositions used with the same verb. Do question 1 with the whole class as an example and then ask students to work in pairs. When going through the answers, point out that in many of these cases, the correct preposition depends on whether it is followed by a person or an object.

5 This may be a sensitive issue with some students or in some cultures, in which case it may be best to omit it.

ANSWERS

Ex. 1

in: result, specialise, confide
for: pay, apologise, apply
with: coincide, compare, contrast
from: benefit, refrain, suffer
on: congratulate, concentrate, insist
to: refer, confess, react
about: boast, worry, learn

Ex. 2
1 one's own
2
1 in 2 on 3 to 4 on 5 by 6 by 7 to
8 in 9 in 10 with 11 from 12 from

Ex. 4
1 of 2 to 3 from 4 on 5 to 6 for

Reading 2: multiple choice (Part 1) p.72

Aims:
- **to practise answering global multiple-choice questions on different text types for CAE Paper 1, Part 1**
- **to practise inferring the meaning of unknown words**

1 This topic may have been discussed in some detail in the lead-in to the grammar exercise above, in which case you may just focus on the question about changes.

2 The multiple-choice questions focus on the writer's overall argument rather than on details. Students skim the three texts and get a general idea of each one. You could set one or two gist questions such as *Which text talks about the problems of living in an extended family?* Then students read the questions for the first text carefully and underline the important words. Point out that the alternatives may contain more than one idea; for example 1A contains the idea of going against convention and the fact that it was unusual, so they need to make sure all the details in the alternative that they choose are correct. Then students read the first text, thinking about the writer's overall purpose, and decide. Check the answers and then ask them to follow the same technique with the second and third texts.

3
1 Ask these questions to the whole class.
2 Ask students to match the words that they already know and then use the contexts to work out the correct meanings of the remaining ones.

ANSWERS

Ex. 1
1 a) extended family b) nuclear family

Ex. 2
1 B 2 C 3 D 4 B 5 A 6 D

Ex. 3
1 l) 2 h) 3 a) 4 b) 5 j) 6 g)
7 c) 8 d) 9 e) 10 k) 11 f) 12 i)

Grammar 2: substitution/ellipsis p.74

Aim:
- **to raise awareness of ellipsis and the use of referencing devices to substitute and avoid repetition**

1
1 Students now focus on the referencing words in the texts and identify what each one refers to.
2 This is best done as a whole-class activity. Students read the texts again and identify which words are elided. Point out the frequent elision of the relative pronoun plus auxiliary *be* in the passive and of repeated verbs when clauses are linked by *and* or *but*.

2 This exercise is best done in pairs. Check the answers by asking different pairs to read out one of the dialogues, pointing out that some of the substitute words, like *not*, can carry the stress, whereas others like *so* are unstressed.

3 Students decide in pairs. Go through the answers and establish that the elided form generally sounds better.

4 Students work in pairs to correct the mistakes. Again check the answers by asking pairs to read out dialogues, with natural stress and intonation.

5 Students work in pairs to improve the text. If they are reluctant to alter the text much, tell them that they must make at least six changes.

6
1 Students work in pairs to identify which response is incorrect.
2 Students now continue with writing one or more similar dialogues of their own. You could give them a specific topic, like plans for their next holiday. They then form pairs or groups and practise reading them with correct stress.

ANSWERS

Ex. 1
1
1 the city where they grew up
2 living in the extended family
3 parents, grandparents and children living together (extended family)
4 problems in relationships with parents and in-laws (*as soon as my grandparents moved in, life got worse for my father*)
5 extended families
6 a way of life that is now largely abandoned in the West
7 families
8 women

2
1 'which has been' 2 'was' 3 'who were'

Ex. 2
1 there **2** so **3** one **4** not **5** It **6** neither
7 that/it

Ex. 3
1 visit her cousin **2** they feel **3** she is leaving
4 to borrow **5** laptop **6** emailed me **7** call me
8 get annoyed

Ex. 4
1 I can't afford *it*/afford *to*.
2 get a new *one*
3 I ~~do~~ *will*.
4 it's Karen's ~~one~~.
5 Do you think *so*?
6 I expect it *will*/I expect *so*

Ex. 5
It is well-documented that relationships between children and their parents fundamentally affect ~~children's~~ *their* behaviour as adults. But now the importance of *these*/*such* relationships ~~between children and their parents~~ is being challenged as new research shows that a child's relationship with its siblings may have a more important effect on ~~a child's~~ *their* future adult behaviour.
Psychologist Francine Klagsbrun says: 'Our relationship with our siblings is unmatchable. ~~Our siblings~~ *They* are there whether we like ~~our siblings~~ *them* or ~~whether we don't like them~~ not. Other relationships change – parents die, friends drift away, marriages break up, but the relationship with siblings carries on and the memories of life that has been shared with ~~our siblings~~ *them* remain with us long after childhood has ended.'

Ex. 6
1 c) **2** a) **3** d)

Listening 2 p.75

Aims:
- **to give practice in understanding speakers' opinions and attitudes and then specific details**
- **to give practice in inferring meaning from context**

1

1, 2 Students can either complete this activity individually and then compare their ideas in pairs, or they can discuss the adjectives in groups from the beginning.
3 These points can be discussed as a brief whole-class activity.

2

1 Students read the four statements and listen to the recording. They discuss their ideas in pairs before checking the answers.
2 Students now listen again and make notes on specific details, then compare the points they have noted in pairs. Then play the recording again, pausing as necessary, to go through the answers. You can also check some vocabulary such as *tomboy*.

3 Follow a similar procedure to the first listening text.

4 Ask students to work together to try to work out the meaning of the highlighted words, or tell each other if they know the words already. You may wish to replay the final section of speaker two before checking the answers.

▶ Recording script p.98

5 Students discuss these questions briefly in pairs before feeding back. You could ask each pair to tell you two similarities and two differences.

ANSWERS

Ex. 2
1
1 disagree 2 agree 3 disagree 4 disagree
2
a) the length of her finger
b) her love of mechanical toys, dirt, etc.

Ex. 3
1
1 agree 2 disagree 3 agree 4 agree
5 disagree
2
a) mixing with the wrong crowd, getting into trouble/fights
b) he loves being thrown up in the air, risk-taker/aggressive/adventurous

Ex. 4
1 main income earner
2 spoil/over-indulge
3 made fun of/persecuted

▶ Photocopiable activity 6 *Birth order quiz* pp.164 and 165

Exam focus

Paper 3 Use of English: key word transformations (Part 5) p.76

Aims:

- **to present a procedure for Paper 3, Part 5 (key word transformations)**
- **to complete an exam-style sentence-transformation exercise**

Exam information

In Paper 3, Part 5 (English in Use), candidates are required to complete eight key word transformation questions. These can test a variety of grammatical and lexical areas. Students will need a considerable amount of practice in this so that they become familiar with the rules of this type of question (use between three and six words and do not change the given word in any way).

Teaching tips and ideas

To give students extra help with the questions in the key word transformation section, provide them with a sheet or OHT with the words of the answers written randomly over it. They put together the missing phrases and insert them into the correct question. To make this more challenging, omit one of the words in each question, which they then have to supply.

Go over the exam information and suggested procedure. Then students work in pairs to complete the transformations. After five minutes, if students are stuck on some of the questions, prompt them by putting gapped versions of some of the answers on the board (e.g. *It is ____ ____ common to live in extended family units in this country nowadays.*).

ANSWERS

1 not so/as common
2 you will have any/much difficulty in
3 came as a (great/big) surprise to
4 make any/much difference to me
5 who talked her out of
6 wasn't I/was I not informed about
7 is going to make a
8 you lend me your umbrella

Vocabulary 2: easily confused words p.77

Aims:

- **to focus on some commonly confused words**
- **to practise obtaining information from the dictionary**

1

1 Students either work in groups of four, allocating a set of words A to D to each person in the group, or each group can discuss one set of words. Once students have considered the difference in meaning of each pair, supply them with a dictionary and ask them to check, looking at the definition, part of speech and example sentences. For those pairs of words where there is a different meaning rather than a grammatical difference, ask them to write down three common collocations for each word.
2 Each group now joins with another to explain the differences they have found.

2 This is best done individually at first. Students can then compare answers in pairs.

ANSWERS

Ex. 2
1 specially 2 worthless 3 lay 4 infer
5 hardly 6 effect 7 check 8 loose
9 principles 10 memories

Writing: competition entry (Part 2) p.78

Aims:

- **to give practice in writing an exam-style competition entry**
- **to present and practise techniques for making writing interesting**

1

1 Students discuss the ideas in pairs. You could ask students to choose the three best ideas, or to rank them in order of effectiveness. After discussing them in pairs, they report to the class.
2 When students have identified the two inappropriate ideas, ask which text these ideas would be suitable for (a report).

2 This gives students an opportunity to see three of the ideas in Exercise 1 in context.

3

1 Ask students to read the question carefully and underline the important words in the bullet points. Then they read the text and tell you which parts of the text answer each bullet point. Then ask about the first and last paragraphs.

2 Students work individually to change the text by introducing substitution or informal language and then compare answers. Ask one or two pairs to read out their changes to the class.

3 Students work individually to identify the spelling mistakes.

4 This could be given as homework or students could write it in class. If you feel students are likely to copy too much of the model text, you could change the task slightly so that students are asked to describe the best friend ever instead of a family member. In this case the second bullet point could read 'explain how they have helped you'.

ANSWERS

Ex. 1

2 c), f) not appropriate, i) is vital

Ex. 2

And what's the point if you can just get money off your parents?
'If you're so independent why don't you get a job?'
This is tricky; score major points

Ex. 3

1
As the bullet points in the question
first paragraph: to introduce the topic
last paragraph: to provide a conclusion and emphasise the argument

2 ~~the problems~~ – them
~~he doesn't do this~~ – But does he do this?
~~young siblings~~ – younger children
~~What he said~~ – Any direct speech here (e.g. *Leave my brother alone!*)

3 personality difficult babysitter bullying selflessness thoughtfulness

UNIT 6 Review p.79

ANSWERS

Ex. 1

1 however **2** ought **3** In **4** with **5** like **6** for **7** under **8** make **9** up **10** them **11** the **12** until **13** is **14** might **15** get

Ex. 2

1 stressed-out **2** dull mundanity **3** quirk **4** get back to them

Ex. 3

1 principal **2** prescription **3** loose **4** priceless **5** implied

UNIT 7 Creative talents

Exam focus

Paper 4 Listening: sentence completion (Part 2) p.80

Aims:
- **to give practice in listening for specific information in a talk**
- **to give practice in predicting missing information**

1 Go through the exam procedure points 1 to 3 with the class and then ask them to read the gapped sentences 1 to 8. Ask them as a class to guess a possible answer to the first gap, using the introductory sentence, the title and thinking of possible collocations with *annual*. Then students talk in pairs and predict possible answers for questions 2 to 8 in the same way. Go through the answers with the class. You might begin by asking which answers are likely to be a job, a number or a percentage and so on.
Go through points 4 to 7 of the exam procedure and then play the recording. Allow students to compare answers in pairs before playing it for the second time. Check the answers as a whole-class activity.

2 This can be done as a brief class discussion.

3 Students listen again and identify the two items of vocabulary. In addition, draw attention to *popping up* in preparation for the vocabulary exercise.

▶ Recording script p.99

ANSWERS

Ex. 1
1 art(s) festival **2** pattern(s) **3** window dresser
4 travel agency **5** eighty percent (80%)
6 glass **7** farmers **8** work(-)shops

Ex. 3
1 foyer **2** the proceeds

Vocabulary 1: phrasal verbs p.81

Aim:
- **to introduce some phrasal verbs and raise awareness of difference in register**

1

1. Students look at the example sentence and comment on the difference in register and increased interest created by using the phrasal verb.
2. Write an example sentence for *pop out* such as *I'm just popping out to the shops* to illustrate the meaning if necessary.

2

1. Students skim read the text and answer the two gist questions.
2. Students complete the text individually by choosing the correct prepositions and then compare answers.
3. When going through the answers with the class, elicit meanings and typical contexts for the phrasal verbs formed by the other prepositions. Students then write example sentences for these, with the aid of dictionaries if necessary.

Teaching tips and ideas

To follow on from the activity of writing example sentences for phrasal verbs or any other lexical items, ask students to work in groups. They read out their sentences to each other, blanking out the target item, which the others have to guess.

3 For the first question just ask students to comment briefly on the text. If your group is relatively academic, however, they may like to discuss the second question on the importance of permanence in art.

ANSWERS

Ex. 1
1
1 *Pop up* has the idea of unexpectedness
2 *Pop out* means to go out quickly and suddenly for a short time. They share the idea of suddenness, unexpectedness.

Ex. 2
2
1 think of 2 bring about 3 come across
4 sought after 5 go for 6 set off
7 worn off 8 go back

Speaking 1: two-way conversation (Part 3) p.81

Aim:

- **to give practice in discussing possibilities, reasons and consequences and in expressing agreement and disagreement**

Exam information

In CAE Paper 5 (speaking), candidates are required in Part 3 to have a conversation together and make a choice from the given alternatives. In Part 4, the examiner asks some more general questions to both candidates on the same topic.

1 Go over the expressions and ask one or two students to say them with natural intonation.

2 Go through the pictures with the class to check that everyone is clear what kind of exhibits are shown in each. Then ask students to talk in pairs and report their decision to you after five to ten minutes. Monitor the students' speaking so that any basic errors can be corrected at the end. You could personalise the activity by asking the students to choose the exhibits for an arts centre in their own town or area.

3 If students are keen on art, then the more general questions here such as 'What kind of art do you like and why?' could work well as a whole-class discussion. Alternatively, ask the students to work in pairs or groups and give one or two of the questions to each pair. Ask them to tell their ideas to the whole class after five or ten minutes.

Reading: multiple choice (Part 3) p.82

Aims:

- **to give practice in reading to understand opinion and argument**
- **to complete an exam-style multiple-choice reading exercise**

1 Discuss these questions briefly as a whole-class activity as a lead-in to the reading exercise.

2

1. Ask students to skim read the text, stopping when they find each of the names A–D and noting down this person's opinion. Check the answers by asking the class to summarise the four opinions.
2. Students now do the exam task. Having skimmed the text, they read the exam questions carefully, think where the answer will be and go back to the text to read for meaning and decide. At the end, they compare answers in pairs before you check with the whole class.
3. Personalise the theme with a class discussion of the opinions from the text.

3

1. Students work individually to find the words and phrases and then compare answers.
2. This could be done individually or as a whole-class exercise.

ANSWERS

Ex. 2

1

A = artform
B = not an artform
C = artform
D = not an artform

2

1 A 2 B 3 D 4 C 5 A 6 B 7 D

Ex. 3

1

1 scruffy 2 intent on 3 emanate 4 animated
5 flaws 6 innocuous 7 breathtaking
8 pilfering 9 pioneer 10 feverish
11 draw a parallel 12 engrossing

2

1 animated 2 intent 3 flaw 4 breathtaking
5 pioneer 6 engrossing

Speaking 2: individual long turn (Part 2) p.84

Aims:

- **to give practice in talking fluently without interruption**
- **to focus on language for comparing and contrasting**

Exam information

In Part 2 of CAE Paper 5 (speaking), candidates are required to talk without interruption for approximately one minute and to compare two photographs. They are expected to give opinions and speculate about what they see, and not just describe the pictures.

1 Students read the task and note that there are two questions they must answer in their talk.

2 Ask the class to decide how they time their answer and emphasise that they should not spend time just describing the pictures.

3

1. Do an example for one of the pictures by brainstorming notes for a few seconds on the board and then ask

students to do the same for the pictures that they choose. Tell them that their handwriting must be legible.
2 Students swap notes and create sentences.

4
1 Pairs talk together about the sentences that they wrote for each other before carrying out the speaking task.
2 Pairs report back on how well they did and how they felt about it.
3 It would probably be a good idea to ask students to form new pairs before repeating the task.

Grammar 1: ways of referring to the future p.85

Aim:
- **to revise and extend students' knowledge of future forms and their uses**

1
1 This exercise tests students' existent knowledge of common future forms. Ask students to discuss the answers in pairs. In those cases where neither form is likely, as in the last two alternatives, ask students to supply the best form. At this stage, they should be familiar with the basic uses of simple present, present continuous, *will* and *going to*, but may be less confident about the use of future continuous and future perfect.
2 This is best done as a whole-class exercise.
3 Students complete this exercise in pairs. Refer students to the grammar section at the back of the coursebook as necessary, or they could use class sets of other grammar reference books if available. Some students may have difficulty in supplying their own sentences instead of copying examples from the grammar reference. If so, you could help them by supplying time phrases for them to include, such as *Next Saturday*, *By 2009*.

2 Ask students how many meanings they know for the word *trunk* and put three on the board (piece of luggage, tree trunk, elephant's trunk). Then ask which meaning they think the word has in the title. Students read the text to see if they were correct.

3 Students now work in pairs to complete the matching exercise. Emphasise that they should use the context in the article to try to choose the correct use.

4
1 Students read the paragraph as a lead-in to the listening exercise and answer the gist question.
2 Ask students to look at the summary. Tell them that a verb in a future form needs to be put into each gap and ask them to discuss in pairs what it might be. Go through the class suggestions, possibly writing up the best ones on the board. Allow for different answers from the ones on the recording (for example, gap three could be *Sarah admits that she is thinking of buying*). Then play the recording once and ask students to tell you the missing phrases or to compare in pairs. Then play the recording a second time pausing after each gap, so that students can write in the correct words.

▶ Recording script p.99

5 Students work in pairs to correct the sentences. Emphasise that the mistakes are with the form; that is, they do not need to choose a different future form but only correct the form of the ones given.

6 Students complete the sentence transformations individually. Then go through the answers with the whole class.

7 Students talk about each of these topics briefly in pairs. Then ask them to report one thing that they learned about their partner to the rest of the class. Correct students' future forms as necessary.

ANSWERS

Ex. 1
1 are you doing 2 opens/is opening
3 'm going 4 are exhibiting
5 I'm going to see/will be seeing
6 will you be going?
2
present continuous – what *are you doing*
present simple – *it opens* on Friday
going to – what *I'm going to see*
future continuous – to understand what *I will be seeing*

Ex. 3
B 2 (is in for)
C 4 (are on the point of finding out)
D 3 (are to)
E 1 (it will soon be possible)

Ex. 4
1 He's only six
2
1 won't have heard of 2 is going to achieve
3 on the point of 4 will have become
5 will be 6 won't be wasting
7 will be sunning

Ex. 5
1 I've never liked art, and so I'm not about to ~~starting~~ *start* going to art galleries now!
2 I hope we *will* have started our new art course this time next month.
3 Work on the new arts centre *is* due to begin on November 15th.

4 Correct
5 Look – those security men have stopped Sue – she won't have ~~been~~ realised that she can't take her camera into the gallery.
6 I was on the point of ~~buy~~ *buying* the picture when I realised now much it cost!
7 Correct
8 Correct

Ex. 6
1 on the point of buying
2 is due to be
3 are unlikely to have heard
4 no intention of staging

Vocabulary 2: words with similar meaning p.87

Aim:
- **to help students to identify the differences between words of similar meaning**

1
1 This is best done in pairs or groups. When they have chosen the odd ones out, check the answers with the whole class, asking them to explain their choices.
2 Students continue in pairs to think of approximate synonyms for the odd words. If you feel this is too difficult with some (e.g. *password*) you could ask them to choose just three.
3 Again in pairs, students now concentrate on the three words which are similar and identify the differences. They can give definitions but tell them that they can also write example sentences or collocations to help explain. You could divide the class into two for this and give word groups 1 to 3 to one half and 4 to 6 to the other. After they have identified the differences, pairs reform with students from the other group and explain them.

2 This exercise can be done either in pairs or individually before class feedback.

ANSWERS

Ex. 1
1 build 2 password 3 indifference
4 conceal 5 disincentive 6 explore

Ex. 2
1 a) examination b) scrutinised
2 a) ambitious b) motivated
3 a) exhibition b) presentation

Use of English: word formation (Part 3) p.88

Exam information

In Paper 3, Part 3, candidates are required to complete a word-formation exercise.

They are given a text with ten gaps, and in each case have to insert a word formed from the given word. The given word must always be changed.

1 Use these questions for a brief class discussion as a lead-in to the word-building exercise. Tell students about anything you or a member of your family has ever collected.

2
1 Students skim read the text to find the two pieces of advice.
2 Ask students to complete the word building either in pairs or individually. If students still find this type of exercise difficult, go through the text with them first and identify what type of word will fit each gap.

ANSWERS

Ex. 2
1 know when to stop buying, use good judgement about what to keep and what to sell on.
2
1 possessions 2 unfashionable 3 attachment
4 undeniably 5 millionaires 6 fascination
7 valuable 8 imaginative 9 judg(e)ment
10 undoubtedly

Grammar 2: verb patterns p.88

Aim:
- **to revise and extend students' knowledge of verbs followed by the gerund and the infinitive, with and without *to***

1
1 Ask students if they have heard of Robert Ripley. If not, use the title as a lead-in by asking students to speculate on the types of things he might have collected before reading the complete text.
2 Students now work in pairs to put the verbs in brackets in either the infinitive or the gerund.

2 Students now create a record of the grammar they have practised by matching each set of verbs to the correct category. With category 1, you may need to check students' understanding of the two meanings. The difference between *remember doing* and *remember to do* in particular may need demonstrating. Emphasise that verb patterns need to form a regular part of students' vocabulary record and that if they meet any new verb followed by the gerund or infinitive, this needs to be indicated in their notes.

3 The rules here provide guidelines for demonstrating which verbs are typically followed by the gerund and which by the infinitive. Ask students to supply some further sentences using the verbs from the categories and point out how these also illustrate the rules.

4 This exercise can either be done in pairs or individually with students checking together afterwards. Do the first one or two together as a class so that it is clear that students need to put the first verb in the correct tense and then the second in either the gerund or infinitive.

5 Discuss these questions briefly with the whole class.

ANSWERS

Ex. 1
2
1 to sell 2 to have 3 to achieve
4 to interfere 5 to work/working
6 to draw/drawing
7 to pursue 8 collecting 9 acquiring 10 filling
11 to wear 12 to drive 13 using
14 communicating 15 being 16 swimming
17 to display 18 doing 19 to make

Ex. 2
a) 4 b) 6 c) 3 d) 1 e) 5 f) 2

Ex. 3
1 forward to the future 2 before

Ex. 4
1 tell you to throw
2 refuse to get rid of
3 enjoy looking back
4 remembers having
5 mind seeing
6 imagined making
7 failed to realise
8 planned to develop
9 regret missing out
10 persuades you to have
11 urge you to reconsider
12 attempting to remove
13 encouraging you to make

Writing: review (Part 2) p.90

Aim:
- **to give practice in writing a review in response to an exam-style writing task**

1 Discuss these questions briefly with the whole class, using the first question to brainstorm the types of things that are reviewed. If you are teaching in an English-speaking environment, you could bring some newspapers or magazines into the classroom and ask students to find reviews of films, books, restaurants and so on. You could ask students to skim read these and then summarise the reviewer's opinion to the class.

2
1 Go through the typical elements of a review with the whole class and then ask students to read the model text. Before establishing which element is missing, ask one or two gist questions to check general understanding, such as whether the reviewer's opinion is positive or negative.
2 Students now read the missing introduction again on page 86. Discuss the questions here with the whole class.
3 This is best done in pairs. When going through the answers, check that students know how formal or informal the expressions are and ask if they know any other informal ways of expressing these ideas.

3 This can be done for homework or in class. Encourage students to swap reviews and read each other's either at the end of the class or in the next lesson.

ANSWERS

Ex. 2
1
1 introduction to catch the reader's interest
2
1 mixture
2 to make the point dramatically/give immediacy
3 with a rhetorical question
4 a) para 2 3 b) conclusion c) para 2
3
1 less than anxious 2 penniless 3 spot on
4 cartoonish 5 catches the eye
6 profoundly moved 7 only time will tell

▶ Photocopiable activity 7 *Reviews* pp.166 and 167

UNIT 7 Review p.91

ANSWERS

Ex. 1

1 headlines **2** acquisition **3** admission
4 exceptionally **5** admiring **6** contemplation
7 uninterrupted **8** solution **9** considerably
10 expectations

Ex. 2

1 due to **2** been collecting **3** will **4** seeing
5 going **6** to enjoy

Ex. 3

1 Edward's grandfather is a professional painter, which has given Edward a headstart.
2 It is very important to have a good illustrator for a children's book.
3 I found the clean lines of the painting both remarkable and fascinating.
4 Many artists struggle to make a living, and some remain almost penniless during their lifetime.
5 I always read the review pages of the newspaper – I find them thought-provoking.
6 My room is full of clutter – I'm always planning to have a tidying session but never get round to it!

UNIT 8 What keeps us going

Listening 1 p.92

Aims:
- **to develop vocabulary for describing personality**
- **to practise listening for gist and main ideas**

1

1 Students read through the job advert. Check their understanding by asking *What is the aim of the job?* and *What does the applicant need to have?*
2 Students work in pairs to choose the words and phrases. You could make the activity more constrained by asking them to choose the three most important.
3 This can be done as a whole-class activity or again students could talk together and think of three characteristics which describe a self-starter.
4 Students now listen to the recording and make notes. At the end ask the group which of the ideas on the board they heard.

2

1 Students complete the questionnaire individually and add up their scores.
2 Before discussing the questionnaire, students listen to the recording again and make notes on the meanings of the scores. Go through the answers with the class, writing the main characteristics on the board. Play the recording a second time if necessary.
3 Students now discuss their scores in pairs. At the end, ask the class as a whole what type of person they would prefer to work with or to employ.

▶ Recording script p.99

ANSWERS

Ex. 1 sample answers

2
sociable able to work independently
has good judgement reliable
willing to follow set procedures imaginative
trustworthy assertive
has common sense obedient courteous
3 someone who can work independently and motivate themselves

Ex. 2

2
12+ Self-starters: like to be in control, look for advice not supervision, don't always follow rules, find new ways of doing things
6–11: moderately independent, manage their own time, need minimal supervision, fit in with accepted methods and procedures
relatively conformist, open to new ideas
0–5: like supervision and clear rules/guidelines, don't question things, get on with what they have to do, keep things running smoothly, very reliable

Vocabulary: three-part phrasal verbs p.92

Aim:
- **to review and practise some three-part phrasal verbs**

1 Instead of asking students to look at the extracts in the book, you could write the two extracts on the board with the two phrasal verbs blanked out. Tell them that there are three missing words in each case and ask students if they can supply them. If they cannot, play the recording again and ask them to tell you to stop when they hear them.

2 Students match the sentence halves, then think of synonyms for the phrasal verbs from the sentences, using the overall context to infer the meanings. Finally they check their answers in the dictionary and note down the correct meaning of the verbs.

3 This can be done as a written exercise, or a strong group may be able to do it orally.

4 Students complete the sentences and then discuss in pairs. Encourage any follow-up questions.

ANSWERS

Ex. 2

1

1 e) 2 c) 3 a) 4 b) 5 f) 6 d)

Ex. 3

1 send off for **2** faced up to
3 come up against **4** get on with
5 cut down on

Use of English 1: word formation (Part 3) p.93

Aims:

- **to focus on the appropriate information to include in a CV**
- **to complete an exam-style word-building exercise**

1 Ask students if they have written their CVs yet and what types of things are typically included in a CV in their countries. Then they discuss the items in pairs and tell the class which they think are the most important.

2 Students complete the word-building exercise, either individually or in pairs.

3 Round off the activity by asking them to choose one piece of advice.

ANSWERS

Ex. 2

1 qualifications **2** application **3** specific
4 vacancy **5** Clarity **6** recruitment
7 irrespective **8** advisable **9** headings
10 accuracy

Exam focus

Paper 1 Reading: gapped text (Part 2) p.94

Aims:

- **to focus on text structure and cohesive devices**
- **to practise a procedure for the gapped text task, Paper 2, Part 2**

1 Ask students to read through the text *My dream job* to gain an overall idea of the story. Set one or two gist questions for this like *Was the novel a success at first?* and *Who was Pru Menon?* Point out any referencing devices at the beginning and ends of paragraphs. These could be pronouns such as *Once **it** was complete* at the beginning of paragraph 3 or linking devices such as *My **next** logistical headache*.
Next students read through the missing paragraphs A to G. They should underline any pronouns which refer to previous paragraphs and any linking devices between paragraphs.
Do the first two gaps together as a class, pointing out how the referencing devices help them choose the correct answer. For example, *what I'd said* in paragraph E refers back to *I told my boss I was resigning and why* in the first paragraph.
Ask students to complete the rest of the exercise individually and then compare answers in pairs. When going through the answers, ask students how they used any referencing devices.

2 Discuss these questions briefly with the whole class.

ANSWERS

1 E **2** B **3** A **4** G **5** D **6** F

Use of English 2: multiple-choice cloze (Part 1) p.96

Aim:

- **to practise a multiple-choice cloze for Paper 3, Part 1**

Exam information

In CAE Paper 3, Part 1, candidates are required to complete a multiple-choice cloze of 12 items. Separate items can test students' knowledge of either lexis or sentence structure.

1 Use one or more of these questions to conduct a brief class discussion.

2 Ask the class as a whole to suggest some possible problems and then skim read the text to see if they are mentioned.

3 Students complete the multiple-choice cloze either individually or in pairs.

4 Discuss the question briefly with the whole class. If you are teaching teenagers, you could ask them to choose which business they would prefer to start and plan how and where they would do it.

ANSWERS

Ex. 3

1 B **2** C **3** A **4** D **5** B **6** A **7** D **8** B
9 C **10** A **11** D **12** B

Grammar 1: direct and reported speech p.96

Aim:
- **to revise reported speech, focusing on backshift and common reporting verbs**

1

1 Use the title of the text to get students to speculate about who might have given the prize and the possible benefits of staff spending less time at work.
2 Students skim the text to check their predictions.

2

1 At this level students should be familiar with the basic tense changes in reported speech. If any students in the group are still uncertain, they can work in pairs.
2 Go through the answers to 2.1 and elicit answers to these grammar questions.

3

1 Students practise converting reported to direct speech. Again this could be done individually or in pairs.
2 Go through the answers to 3.1 and complete the rules, pointing out that there is no tense change in 2 as it is a general statement which is still true now, and the reporting verb is in the present (*she believes*).

Watch Out! *say* and *tell*
The use of the verbs *say* and *tell* is frequently confused. Elicit the rule that *say* is usually followed by a clause and never has a person as its object whereas *tell* usually does. Brainstorm some other common collocations with *tell* such as *tell a story, tell a lie*.

4 Students complete the exercise individually or in pairs.

5 This can be done in open pairs. Give students a few moments to think and then ask individual students to change the utterances to direct speech. At the end, play the recording to reinforce the correct answers.

6

1 Ask students to listen to the recording and make notes. Then individually they write a summary in reported speech. Check the answers by playing the recording again and eliciting sentences in reported speech.
2 Students now carry out a similar activity in pairs. If you wish to make this a spoken activity, each student could also make notes on his/her partner's speech and then report his/her ideas to the class using reported speech.

▶ Recording script p.100

ANSWERS

Ex. 2

1 having kids made him realise (that) there was more to life than work
2 his staff would be happier if they could see their children more
3 they were/had felt uneasy about it at first, but they soon started to appreciate it
4 if staff are happy, they work better, are more loyal and less likely to leave the company
5 the office is/was open longer than before because of flexible working hours

2
1 sentence 2
2 sentence 4
3 sentences 3 and 5
a) would, could, past
b) clear
c) present

Ex. 3

1
1 I awarded the prize last month for the example Ian's forward thinking has given to others.
2 I believe many people want to break out of the long-hours culture.
3 I accept Ian's company has benefited from higher productivity and greater flexibility.
4 Lower staff turnover should help to convince other organisations that this is the way forward.

2
a) doesn't/needn't b) *should*

Watch Out! *say* and *tell*
1 a)
2 a)

Ex. 4
1 told, said **2** said, told **3** told, said
4 said, told

Ex. 5
1 I was wrong to get angry.
2 I earned more than ever last year.
3 I've never met him before.
4 Please think about what you are doing!
5 I will work harder next week.
6 I believe that overall performance will improve if we give bonuses to our staff.

Ex. 6 Sample answer
She said that she had always wanted to work for herself, but she had never thought it would happen. Then someone asked her to write a story for the local magazine. Because she found it/had found it really easy, she decided to write another one. One thing led/had led to another and now she is writing full time, and she loves it!

Listening 2: multiple choice (Part 3) p.98

Aim:

- **to practise an exam-style multiple-choice listening task**

Exam information

In Part 3 of CAE Paper 4, Listening test candidates are required to listen to a passage and answer six multiple-choice questions. They will hear the passage twice. The questions focus on understanding the speaker's attitude and opinions.

1 You may need to pre-teach the term *soap* or *soap opera* and give an example from the students' own culture. Then use the questions to conduct a brief class discussion as a lead-in for the listening activity.

2

1 Ask students to look at the six multiple-choice questions. Read question 1 together and decide which are the important words to underline for each option, for example:

   A one of her friends was already working on the programme.

   Students then read the rest of the questions and underline the important words in the same way. Ask in which questions the options are a reason for the action in the stem (1, 3 and 5).

2 Students do the vocabulary exercise individually.

3 Students now listen and choose the correct answers. After the first listening, they can compare in pairs before they listen to the recording again. Finally, go through the recording pausing as necessary to discuss the answers. In each case, establish why the other options are incorrect.
Point out the importance of the relationship between the stem and the options. As a fact on its own, A is correct but we are looking for the reason why she accepted the part.

▶ Recording script p.100

4, **5** Use this as a brief discussion to round off the activity.

ANSWERS

Ex. 2

2

1 financing 2 flattered 3 lifelong ambition
4 settled into 5 falling out 6 announced
7 letting people down 8 became anxious

Ex. 3

1 C 2 B 3 A 4 C 5 B 6 D

Speaking: comparing (Part 2) p.99

Aims:

- **to practise talking fluently without interruption**
- **to practise using the language of comparing and contrasting**

1 Allot one or two of the places to each pair and ask them to brainstorm advantages and disadvantages. Then they report their ideas to the class.

2 Ask students as a class what kind of environment they like to work or study in (with others or alone, in silence or with background noise or music, with their own or shared space, etc.). Then give two photographs to each student and ask them to talk to a partner for about one minute, comparing the two environments.

3 Discuss these questions briefly with the whole class.

Teaching tips and ideas

For further work on Paper 5, Part 3, if possible, ask a native speaker before the lesson to compare two of the photographs for a minute and record his or her response. Students then listen to this after they have completed the task to identify the two photos and compare his/her ideas with their own.

Grammar 2: reporting words p.99

Aims:

- **to focus on using a range of structures with different reporting verbs**
- **to focus on the need to change certain time phrases in indirect speech**

1 Ask students first to match each of the sentences to the correct reported statement, according to the meaning of the reporting verb. Then tell them to finish any of the sentences which are still incomplete, making any necessary changes to the verb or time phrase. When going through the answers, build up a list of how the time phrases may need to change on the board.

2 Students now record the structures used with the different reporting verbs in Exercise 1 by inserting them into the table. They work in pairs to add the extra verbs in the box.

Watch Out! *suggest*
It is a common mistake to use *suggest* with an infinitive as in sentence c), partly because of the structure with other verbs with similar meanings like *advise*.

3 Students complete this exercise in pairs or individually.

4 In pairs students match the verbs and then use them to rewrite the sentences. Tell them to refer to the table in 2 if they are unsure.

ANSWERS

Ex. 1

1 b) **2** f) **3** d) **4** g) **5** e) **6** c) **7** i) **8** j) **9** a) **10** h) **11** k

Ex. 2

Verb + object + prep + *ing*	**Verb + inf**	**Verb + prep + *ing***
congratulate (on) accuse (of) thank (for)	offer refuse threaten invite ask agree intend promise propose	apologise (for)
Verb + *that*	**Verb + object + inf**	**Verb + *ing***
announce remark claim explain confirm suggest propose	order advise remind command	admit suggest

Ex. 3

1 promised
   She announced that she would attend the meeting.
2 expected
   He insisted on receiving the invoice before he paid for the goods.
3 agreed
   The manager confirmed that he accepted the policy decision.
4 ordered
   She demanded that he arrive on time.
5 offered
   The employees suggested taking a small pay cut.
6 congratulated
   The manager thanked me for changing the work ethic in the office.

Ex. 4

1 He reminded me to send the email today/that day.
2 He advised me to resign immediately.
3 He congratulated me on getting promoted.
4 He confirmed that he wasn't going to do the training course the following year.
5 He complained about the food in the canteen.
6 He asked Michael to help him write his proposal because he couldn't do it.
7 He announced that the managing director would leave the company the following week.
8 He apologised for being late and explained that there had been a problem with the train.

▶ Photocopiable activity 8A *Risk transformations* p.168

Writing: proposal (Part 1) p.101

Aim:

- **to practise writing a proposal in response to an exam-style writing task**

1 Students look at the two plans and decide as a class.

2

1 Students decide in pairs or individually which statements are true. Point out that linking words are an essential part of any text.
2 Complete the two sentences together.

3

1 First ask students to read the whole task carefully. Check their understanding by asking them to tell you the overall topic, and what could be included from the comments and Internet research.
2 Students now read the example proposal.

4 Students look at the proposal again. Establish that the information from the survey is factual and therefore comes under the *background information* heading whereas the Internet research is the basis for the suggestions.

5

1 Students work in pairs to find the expressions in the proposal.
2 Ask students to look for the modal verbs. Establish that *should* is used frequently in the suggestions section whereas the background information is factual and so uses present simple tense.

6 Students work individually to add a final key feature. They should look back at the Internet research for ideas. Then pairs read their bullet points to each other.

7 Students write their own proposals either in class or for homework. With a strong class, you could change the task a little so that they have more opportunity to use their own ideas, such as asking them to write a proposal for an ideal study area in their school or college instead of an ideal workplace.

ANSWERS

Ex. 1
A proposal **B** report

Ex. 2
1
It includes suggestions and recommendations.
It may use headings or bullet points if appropriate.
It proposes a new idea and tries to persuade the reader of its value.
2
1 proposal 2 report

Ex. 4
comments from survey – background information, it's based on actual research
research from the Internet – suggestions, it provides the basis for the suggestions

Ex. 5
1
1 dislike
2 is a particular issue
3 maintain motivation
4 claustrophobic
5 have concerns over
2
'should' – it's used for recommendations

Ex. 6
noise – the phone has been omitted – mobile phones must be on vibrate

▶ Photocopiable activity 8B *Compound nouns dice game* p.169

UNIT 8 Review p.103

ANSWERS

Ex. 1
1 be taken into
2 is no/little/hardly any point (in) (your) attending
3 admitted (that) his interview had not
4 the exclusive use of
5 accused her of leaving
6 is essential to have a CV
7 apologised for breaking Brenda's
8 wished (that) she had gone to

Ex. 2
1 work **2** charge **3** range **4** point **5** loss
6 limit

Ex. 3
1 deny **2** advise **3** accuse **4** admit **5** refuse
6 agree

UNIT 9 On the road

Speaking 1: choosing an image (Parts 3 and 4) p.104

Aim:

- **to practise expressing opinions and negotiating a decision**

1 You could begin by constructing a word diagram on the board for *inspire* (*inspiring, uninspiring, inspiration, inspirational*). Then students think for a moment and choose something they have found inspiring to discuss with a partner.

2 Students discuss the images for the poster and make their choice. Finally, they tell the whole class which one they decided on and why.

3 The questions can be discussed with the whole class or students could talk about them in pairs.

Grammar 1: review of narrative tenses p.105

Aim:

- **to review the use of present and past tenses in narrative and provide controlled practice**

1 Students work individually to choose the correct tenses and match to the statements. Allow them to check in pairs before going through the answers.

2 Students complete the tense exercise either individually or in pairs.

3 Ask the students' opinion on the story in Exercise 2. Then ask them to think of a similar story of a holiday or journey where there were many mishaps or events. It can be either personal or something that happened to someone they know. Give them a few moments to prepare the story. Then they tell the stories to each other in pairs or small groups. Round off the exercise, if possible, by telling a story of your own. Students then write their story as a paragraph. To encourage the use of a range of structures, you could tell them that they must include at least two examples of the past perfect and put the following sentence frames on the board:

Once I/we ... , I/we ....
After I/we ..., I/we ...

ANSWERS

Ex. 1
1 are sitting c)
2 went out, never saw g)
3 was travelling a)
4 had already left b)
5 've finished e)
6 read d)
7 'm visiting f)
8 've been staying h)
9 'd been living i)

Ex. 2
1 has become 2 had collapsed 3 persuaded
4 had arrived 5 were obscuring 6 had come
7 knew 8 was sucking
9 have never felt 10 had been raining
11 arrived 12 had picked up
13 have not climbed/have not been climbing
14 are you climbing/will you be climbing

Reading: multiple matching (Part 4) p.106

Aim:

- **to give practice in understanding specific information and opinion**

1 Ask students to look at the headline of the article and, either as a whole class or in pairs, suggest some ways in which tourists could make friends with or annoy local people. If this is not a sensitive subject, you could ask for some examples of annoying behaviour by tourists in their own countries and what tourists need to know to avoid making a bad impression.

2 Pairs now discuss these questions together. In a multilingual class, the questions about gifts and souvenirs provide a good opportunity for discussion and comparison of the types of objects which are typical of different countries, and so you might focus especially on this.

3 Students skim the text quickly and match the topics with the sections.

4 Students work individually to complete the multiple-matching task. Before they begin, remind them of the techniques that they need to use. They should read the questions first, then section A of the text. At this point they should go back to the questions again, read down and mark those which correspond to this section. Then they should follow the same procedure with the remaining sections. Stronger classes may be able to do this without the initial reading of the questions. If any questions seem to refer to more than one section, they should go back to both sections of text and check the relevant sentences carefully.
Remind students that situations which are described in specific detail in the text are often referred to in general terms in the questions; for example travelling *in a relatively uncomfortable way* refers to taking a third-class place on a train.
You might also tell students before they begin that the phrase *the beautiful game* refers to football. (The expression originated from the Brazilian football player Pelé, who published an autobiography entitled *My Life and the Beautiful Game*.)

5 Students work in pairs to complete the collocations. They should leave blank any that they do not know and then go back to the reading text to find the correct collocations. They can also check their answers in this way.

6 Students may already have discussed the behaviour of foreign tourists in some depth in the lead-in to the reading exercise. If so, you can simply ask them their opinion on the advice in the text or ask them to pick out the most important piece.

ANSWERS

Ex. 4
1 E **2** C **3** A **4** E **5** C **6** D **7** E **8** B
9 D **10** A **11** D **12** A **13** E **14** B **15** A

Ex. 5
1 break **2** get **3** make **4** set **5** take **6** go
7 meet **8** give **9** stand **10** lose

▶ Photocopiable activity 9 *Travel and transport idioms* pp. 170 and 171

Vocabulary: dependent prepositions – adjectives and nouns p.108

Aim:
- **to revise and extend students' knowledge of prepositions after adjectives and nouns**

1
1 Students work in pairs to decide on the correct prepositions. Then go through the answers with the whole class. You could also provide dictionaries for students to check their answers.
2 This question is best done as a whole-class activity. It provides an opportunity for students to extend the exercise by brainstorming some common noun or gerund collocations of these adjectives and prepositions.

2 These questions allow students to practise some of the adjective and preposition combinations in a freer context. You could extend the exercise by asking students to invent some further questions using the adjectives and prepositions, which they ask each other in closed or open pairs.

3, **4** Introduce the exercise by asking students to supply the nouns formed from some of the adjectives in Exercise 1. Then do question 1 with the whole class as an example before asking students to complete the rest of the exercise in pairs.

5 This question can be discussed briefly as a whole-class activity. However, one obvious reason for placing restrictions on travel is the effect it has on the environment and the contribution of emissions from cars and planes to global warming. If the class are interested in this topic, you could ask them to work in teams and each present a short proposal on how these emissions could be cut in their home town or country.

ANSWERS

Ex. 1
1
1 at 2 about 3 from 4 to 5 of 6 by
7 to 8 with
2 either

Ex. 3
1 separation from **2** truth of
3 fascination with **4** nervousness about

Ex. 4
1 similarities between **2** concern for
3 differences between **4** pleasure from

Listening: sentence completion (Part 2) p.109

Aims:

- **to give practice in predicting and listening for specific information**
- **to focus on checking for common mistakes**

1

1 Ask students to speculate on the topic of the listening text by looking at the pictures from the website and establish that this is a gastronomic tour by bicycle.
2 Students work in pairs to brainstorm equipment needed and possible problems. If you wish, you could give the first task to one half of the class and the second task to the other. Then they should feed back to you to create two lists on the board. If students cannot think of many ideas at first, encourage them to think of any cycling journeys they have made.

2

1 Students now listen to the recording and answer these three initial questions. The answers appear early on in the recording so you may choose to play just as far as the phrase *leg of his journey* at this stage.
2 Before listening to the complete sentences, ask students to predict likely words to complete the sentences, such as what might cause problems in the gears of a bicycle or what he might have lost. Point out that, as in the reading exercise above, the questions may use a more general phrase such as *an American food item* to describe something that is mentioned specifically in the text. You could also draw students' attention to the use of the superlative in questions 5 and 8, which mean that he may have had a number of problems or tried a number of strange dishes, but they need to pick out the one which is the greatest or strangest.
After students have completed the sentences, they check answers in pairs.

▶ Recording script p.101

3 This question anticipates some common mistakes that students may make in an exercise of this type. Students work in pairs to identify why each answer would not gain them a mark in the exam. Then conduct whole-class feedback and build up a list on the board of the types of errors they need to watch for. This probably works best as a series of check questions, for example:
Is the singular/plural correct?
Is the spelling correct?
Does the word or phrase fit the gap exactly?

4 These questions can be discussed with the whole class briefly to round off the exercise, or you may choose to use them to form the basis of a more extended discussion task. Students work in groups to plan a route or itinerary for Tom for a set number of weeks to explore and sample part of the traditional cuisine of their own country. They then give a short presentation of the route they have chosen to the rest of the class. This of course would work best in a monolingual class or a class where students can work for once with classmates from their own country.

ANSWERS

Ex. 2
1 graphic designer 2 sponsors/sponsorship
3 dust 4 bell 5 punctures 6 gloves
7 fruit pies 8 soup

Ex. 3
1 Tom is a designer – not a design: think about the kind of information you are looking for and check your spelling!
4 Only one bell: check your grammar!
5 Hills are a problem, but not the biggest one: listen carefully!
6 It doesn't fit grammatically: read the whole sentence carefully and don't repeat the information that's already there!
7 Roadside diners are not a food item: Read the sentence carefully – make sure you know what information you're looking for!
8 This is the taste, not the type of dish: don't forget to read what comes after the gap!

Grammar 2: emphasis (cleft sentences with *what*) p.110

Aim:

- **to present or revise cleft sentence structures and to provide controlled practice**

1 Students look at the example from the listening and say what the effect is of using the cleft structure.

2 Ask students to look at the further examples as a class and tell you what part is being emphasised.

3

1 Students work individually or in pairs to rewrite the sentences. When going over the answers, ask students to say the cleft sentences with natural sentence stress and intonation.
2 This could be done as a written exercise or orally.
3 Ask the class to tell you the best way to complete the statement.

4

1 You could extend this exercise by asking students to complete each of the sentences first.

2 Students listen to the recording and answer the gist questions.

▶ Recording script p.101

3, 4 Students choose one of the topics and prepare their talks. You could allow them to change *last week* in topic 2 to a different time reference if necessary.

ANSWERS

Ex. 1
To highlight the importance of the following information – to 'front' it.

Ex. 2
1 the object
2 the verb or event
3 a whole sentence

Ex. 3
1
1 What she did was learn Italian so that she could speak to people when she went to Rome on holiday.
2 What annoys me most is people who are always late.
3 What made him determined to go back was that he couldn't visit Iguazu when he went to Brazil the first time.
4 What he did last year was go on a course to become a flight attendant.
5 What I'd really like to do is have a holiday in Antarctica/What I'd really like to have is a holiday in Antarctica.
6 What she told her boss was how she felt about the restructuring of the company.

2
1 He really loves travelling by plane.
2 I like to take photographs of every place I visit.
3 I use the Internet to stay in touch when I travel.
4 I am enjoying the chance to travel abroad for my work.
5 I was furious about the delay, and so I complained directly to the airline.
6 To my amazement, the airline refunded all my money!

3
informal speech

Speaking 2: individual long turn (Part 2) p.111

Aims:
- **to give practice in comparing situations, giving opinions and speculating**
- **to give practice in speaking without interruption for one minute**

1 Ask students to look at the pictures and invite some brief comments on the differences between the types of people and the way they are probably feeling.

2 Go through the example with the students. They then put the language from the example into the correct part of the table. Ask them to suggest any other possible expressions that could be used for these functions.

3 Students now listen and note down any additional expressions that they hear. If necessary, play the recording again, pausing before each relevant expression and building up an additional list on the board.

▶ Recording script p.102

4 Students now do the speaking task in pairs. Give a time limit of one minute for each student.

ANSWERS

Ex. 2
What you think/speculating: seem to, appear to, what I think is
Comparing and contrasting: in the same way, this time, whereas
Qualifying what you say: a bit more
Giving reasons/explaining: for that reason

Exam focus
Paper 3 Use of English: open cloze (Part 2) p.112

Aim:
- **to present and practise a procedure and techniques for the open-cloze task**

Exam information

In CAE Paper 3, Part 2, students complete an open cloze. This can test a variety of lexical and grammatical areas, including prepositions, collocations and link words.

1 Ask students to skim the *holiday snaps* text and tell you what the writer's general point is about taking photos (that taking photos may discourage us from looking carefully at something). Then go over the suggested procedure with the class.
Students work in pairs or individually to fill the gaps.
At the checking stage, if students have written different answers, you could write them all on the board and ask the class to identify which answers are not possible and why. Point out that students need to be especially careful in those cases when finding the right word for the gap depends on their awareness of the structure of the whole sentence, not just the words on either side of it. This is true of question 12 and question 14.

2 These questions can form the basis of a whole-class discussion to round off the exercise. You could also ask students if they have any other preferred ways of remembering or creating a record of places that they have visited (e.g. collecting postcards or writing a description in a diary) and which are best.

ANSWERS

Ex. 1
1 ourselves **2** out **3** the **4** more **5** do
6 Even **7** one **8** in **9** against **10** instead
11 however **12** will **13** what
14 (al)though/while/whilst **15** by

Writing: competition entry (Part 2) p.113

Aims:
- **to practise writing an article**
- **to focus on responding to a competition stimulus**

1 Students look at the photo and comment briefly about the message that it seems to give about travel. Ask what experiences they have had of stress and delays.

2 Students now read the instructions for the competition entry carefully and choose the most suitable answers for the two multiple-choice questions. They compare their ideas in pairs.

3
1, 2 Students turn to page 208 which gives guidance on writing articles. As a whole class, ask them to identify the four aims for an article and the best four ways of achieving them.

4
1. Students now work in pairs to decide on some advice. You could ask them to write down the best three pieces they think of to tell the class.
2. Students listen to the advice on the recording and compare the ideas with theirs.
3. Students listen a second time and make notes. Build up the advice on the board.

▶ Recording script p.102

5 Students now read the example article and discuss the questions in pairs. Conduct a feedback session, focusing especially on the examples that students have found.

6
1. The planning stage of this task is probably best done in class. Students plan their paragraphs including the introduction and conclusion in note form and then compare their ideas in pairs. If students have difficulty in starting, you may choose to brainstorm possible points as a whole-class activity and then ask them to work alone to plan the paragraphs. This may also be a good opportunity to compare and discuss briefly in what order they usually plan their content. For example, some people typically prefer to plan the introduction last.
2. The actual writing can be done in class as well or for homework. Stronger classes who may need less guidance should probably choose a different photo from the one on page 113 or even one of their own that they could bring to class. Point out that students should use the grammar checklist or the checklist in Exercise 3 before they hand their work in. You could ask them to swap articles next lesson so that a classmate compares their work with the two checklists.

ANSWERS

Ex. 2
1 B **2** A

Ex. 3
1 a) b) c) e) **2** a) b) d) e)

Ex. 4
2
start with the photo
give a bit more detail
write down your ideas clearly
use interesting vocabulary
give your own opinion
try to win

3
give a bit more detail on what you can see
write down your ideas clearly
use lots of interesting vocabulary
do not give too much unnecessary detail
give your own opinion
try to win

Ex. 5

1 To introduce the topic and engage the reader
2 Yes
5 First, then, all in all
6 Rhetorical questions, speaking directly to the reader with exclamations
7 Informal – speaking directly to the reader, e.g. *'But it's not all good news on the personal front'; 'Of course, there is another side to the coin – I can't pretend it's all bad news.'*

UNIT 9 Review p.115

ANSWERS

Ex. 1

1 attendant **2** offensive **3** inappropriate
4 employee **5** procedure **6** boarding
7 sympathised **8** patience **9** behaviour
10 inconvenience

Ex. 2

1 to provide readers with
2 as if they have (some)
3 that/which annoys me most is
4 a slight fall/reduction/decrease

Ex. 3

1 unnecessary **2** uneventful **3** disrespectful
4 inconvenient **5** impractical **6** inconsiderate
7 unrelated **8** misbehaviour

UNIT 10 Close to nature

Listening 1 p.116

Aim:
- **to practise listening for opinion and attitude**

1

1 Begin by writing the topic *environmental issues* on the board and asking students what individuals can do to help the environment. Ask them to name one thing they already do and one further thing that they could do. Students look at the multiple-choice questions. Play the recording and ask them to decide in pairs after each extract which option is closest to the speaker's attitude. After this, play it again and highlight any important lexical items like *bury your head in the sand*.
2 Discuss briefly with the class which speaker they agree with.

▶ Recording script p.102

2

1 Students read and complete the quiz individually and add up their scores. Then they talk in pairs and compare answers.
2 Pairs look at the quiz questions again and discuss the first three questions in relation to the issues in each one such as packaging. Discuss the fourth question briefly with the whole class.
3 Students work in pairs to write three more questions which they ask and answer in groups. Alternatively, each pair could write just one question and then do a mingling activity where they walk around asking the other pairs their question for a mini classroom survey.

ANSWERS

Ex. 1
1 C **2** B

Grammar 1: countable/uncountable nouns p.117

Aims:
- **to revise the grammar of countable and uncountable nouns**
- **to give practice in completing a gapped sentence exercise**

Exam information

In Paper 3, Part 4, there are five sets of three sentences. Students have to think of one word which can be inserted into all three sentences in each group. The word must be exactly the same for all three sentences.

1 Students discuss the sentences in pairs and decide which of the two alternatives is possible or more likely. When checking the answers, point out examples of words which have no countable form such as *advice* and words which can be used in both forms with a change of meaning such as *space*.
You might also point out differences here with students' own languages as there are a number of words which are uncountable in English but countable in many other languages like *information*.

2 Students now work in pairs to discuss the words here.

3 A switch from countable to uncountable use is one of the ways in which the use of the words in this exercise type may vary between the three sentences. When going through the answers, point out that *idea* in question 5 provides an example of this.

ANSWERS

Ex. 1
1 space **2** Iron **3** some advice
4 much news **5** a coffee
6 an amazing time **7** excellent research
8 a hair (hair on the head or body is usually uncountable, a single hair (e.g. in food) can be countable)
9 additional information **10** Travel

Ex. 2
1 both (countable if it means 'a cup of coffee')
2 both (countable if referring to a specific hope)
3 uncountable
4 countable
5 countable
6 uncountable (except in the sense of an influential country or superpower)
7 both (countable if it means a person with authority)

Ex. 3
1 eye 2 chances 3 condition 4 strength
5 idea 6 interest

▶ Photocopiable activity 10A *Link words: Pickles and the world cup* p.172
▶ Photocopiable activity 10B *Commas and colons* p.173

Reading: gapped text (Part 2) p.118

Aim:
- **to complete an exam-style gapped text**

1
1 Introduce the topic of the reading text by writing the word *extinction* on the board and eliciting the adjective *extinct* and some collocations such as *face extinction*. Point out that there is no single corresponding verb. Then students work in pairs and brainstorm animals which are in danger of extinction and the main reasons.
Students then look at the picture in the book and discuss together which they think are the correct figures.
2 These questions can either be discussed in pairs or as a class.

2 Introduce the reading text by asking students to look at the photo to identify the species and the part of the world. Then ask them to read the title and headline to check.
Go over the procedure for this type of task with students (skim the text first, read the list of missing paragraphs carefully and then match each one one at a time, paying attention to the reference words and phrases). Students do the task individually and then compare their answers in pairs.

3 These questions can form the basis of a class discussion.

4 Ask students to work in pairs or individually to match the verbs and objects. They check their answers by finding the collocations in the reading text.

ANSWERS

Ex. 1
1 c) 2 e) 3 a) 4 d) 5 b)

Ex. 2
1 B 2 F 3 A 4 G 5 E 6 D

Ex. 4
1 d) 2 h) 3 a) 4 g) 5 f) 6 e) 7 c) 8 b)

Exam focus
Paper 4 Listening: multiple choice (Part 3) p.120

Aim:
- **for students to complete an exam-style listening exercise**

1 First go over the exam information and suggested procedure with the students. Then ask students to look at the exam task and read the first multiple-choice question. Ask them to suggest what the most important words are in the stem. Then they look at the four alternatives and again suggest what the most important words are.
Students then read the rest of the questions, again underlining the important words. Ask them if they can suggest any phrases they should listen for in each case which will indicate that the answer to that particular question is coming, such as 'arthropods' for question 1. These will often be the same as the words they underlined.
Play the recording twice and ask students to choose the correct answers, following steps 3 and 4 of the suggested procedure. They compare answers in pairs before checking as a whole-class activity.

2 Use these questions for a brief class discussion on the content of the listening.

▶ Recording script p.102

ANSWERS
1 A 2 D 3 C 4 B 5 A 6 B

Use of English: open cloze (Part 2) p.121

Aim:

- **for students to complete an exam-style open cloze exercise**

Teaching tips and ideas

To revise a previously completed cloze exercise, you can use the oral cloze technique. Take the text that students completed in a previous lesson and read it aloud to them blanking out the answers, which they have to remember and supply as a class.

1

1 Ask students to look at the photo of the police dog. Draw their attention to the difference in salaries and length of career. Then ask the class to suggest answers to questions 1 and 2.
2 Students skim the text to see if their questions are answered.

2 Students now work either in pairs or individually to fill the gaps. Remind them of the suggested procedure on page 112.

3 Students complete the vocabulary exercise individually or in pairs. This could be followed up by a discussion (with the whole class or in pairs) on issues relating to the use and training of animals (e.g. *Do you think it right to train animals like this, Do you think Keela enjoys her work, What else might Keela be useful for?)*

ANSWERS

1 being **2** put/placed **3** such **4** up **5** only **6** what **7** is **8** how **9** whose **10** as **11** where **12** get/become **13** in **14** This **15** which

Ex. 3

1 an asset
2 to have the perfect temperament
3 pinpoint
4 unique talents
5 much in demand

Grammar 2: Introductory *It* p.122

Aim:

- **to raise students' awareness of the use of *it* as a preparatory subject and object**

1

1 Look at the example with students and establish that the use of the *it* gives added emphasis to *the heavy rain*. Ask students to try saying the second sentence with the correct stress and intonation. Then they look at the pairs of sentences a and b, underlining the emphasised information.
2 Check the answers to these questions with the whole class.

2 Students work in pairs to transform the five sentences. Check the answers by asking individual students to read them out with natural stress and intonation.

Watch Out!

This question is particularly relevant to students who speak pro-drop languages such as Spanish or Italian.

3 Ask students to look at the example sentences and tell you what the object of the verb *find* is in the second. Point out that the clause in italics is the real object but *it* also functions as a preparatory object. Students then work in pairs to insert the preparatory *it* in sentences 1 to 7.

4 Students will already be familiar with these structures. They work pairs to transform them. Draw attention to the collocation *common knowledge*.

5 Students again work in pairs to correct some of these sentences. When going over the answers, check that they understand that *owe* in sentence 7 does not refer to money but simply means that we deserve honesty.

6 This exercise personalises the grammar of this section. You might wish to delimit it by telling students that all sentences must be about the natural world. You may need to preteach the meaning of the phrase *I take it that*.

ANSWERS

Ex. 1

1

1 a) John, not James b) Saturday, not Sunday
2 a) how important pets are b) what some people will do
3 a) to be aware of environmental issues
b) to find environmental issues on the front page of newspapers

2

1: 3 2: 2 3: 1

Ex. 2

1 It is frightening how easy it is for natural habitats to disappear.
2 It is unlikely that most people feel indifferent to the fate of some species of animals.
3 It was thrilling for us to have seen a condor in the wild.
4 It doesn't really matter when you return my book.
5 It was confirmed yesterday that the conference will take place next month.

Ex. 3

1 She thought *it* was strange that he hadn't contacted her.
2 His headache made *it* difficult for him to concentrate.
3 She thought *it* had been a mistake not to sign the contract immediately.
4 I found *it* exciting that I was asked to take part.
5 His behaviour made *it* impossible for me to continue to work on his project.
6 I hate *it* when he shouts like that.
7 I love *it* when they have fireworks at a party.

Ex. 4

1 It is thought that global warming is caused by human activity.
2 It appears that world temperatures are rising.
3 It is common knowledge that we should recycle wherever possible.
4 It seems that there are many species on the verge of extinction.

Ex. 5

1 I cannot bear ~~it~~ to see people being cruel to animals.
2 ✓
3 ✓
4 He made *it* obvious to everyone that he was not going to get involved in the project.
5 I'll leave it to you to choose ~~it~~ the film we watch.
6 I knew ~~it~~ that they were unhappy about the plan.
7 ✓
8 ✓

Speaking: sounding interested p.123

Aims:

- **to focus on sounding engaged and interested in a conversation using intonation**
- **to practise using phrases to express and invite opinions**

1

1 Introduce the topic by asking students if zoos are a good way to save animals from extinction but do not let this develop into a full discussion at this stage. Then students listen to the two recordings separately and answer the questions.
2 Students now discuss these statements in pairs or groups. After about five minutes, ask the whole class to feed back on their discussion.
The answer to the final bullet point, *'you should talk a lot'* may well depend on how talkative your students tend to be normally. If you are dealing with students who tend to be rather quiet and reticent, you may need to emphasise that they need to overcome this in the exam; on the other hand, if they are usually fluent and talkative, you may need to warn them against dominating the conversation in Parts 3 and 4.
As well as intonation, it may be worthwhile raising the issue of body language. The advice you give will depend on the culture of your students. With some students, you may need to emphasise the importance of making eye contact and using posture to demonstrate that you are listening whereas students from other cultures may need to tone down their gestures and avoid any which are not commonly used in English.

▶ Recording script p.103

2 Students now discuss which of the phrases here are encouraging in a conversation. This depends not only on the words but on the intonation. Demonstrate the same phrase, such as *Do you think so?,* said in both an engaging and an offputting way.

3 Begin by choosing one student to have the conversation with you and demonstrating how to sound bored and then interested. Then ask students to work in pairs to discuss the question, following the instructions.

4 Students discuss these questions in pairs.

ANSWERS

Ex. 1

1

1 the man 2 the woman
3 intonation, vocabulary
4 Lack of interest has a negative effect on the listener, and doesn't contribute to turn-taking. Sounding involved encourages the listener and creates a positive impression.

2

it is a good idea to ask them for their opinion; it is important to answer using more than one or two words; intonation is important to give a good impression

Ex. 2

Engaging: That's really interesting – tell me more; No, I don't really agree – but what I think is …; I totally agree – and what's more …; A good point – it's absolutely true that …
Offputting: Do you think so?; I suppose so.

Vocabulary: phrasal verbs and compound nouns p.124

Aims:

- **to introduce some phrasal verbs and compound nouns with the particles *up* and *down***
- **to focus on the meanings associated with these particles**

1

1 Students insert the correct verbs into the gaps and then compare answers in pairs.
2 As a class, they decide which of the meanings describes the use of that particle in each text.

2 Students now work in pairs to fill in the correct prepositions. For any expressions which are new, encourage them to try to work out the correct particle by referring to the meanings in 1.2.

3 Complete this as a whole-class activity.

4 Students work individually or in pairs to form the correct compound for each sentence. Tell them that sometimes the particle comes before and sometimes after the verb, or may do both with a change of meaning as in *set up* and *upset*. When checking the answers, point out the use of hyphens for some compounds.

ANSWERS

Ex. 1

1

1 keep, catch, made
2 cut, calm, narrowed
3 tidy, freshening, do
4 speak, speeds, livening
5 closed, track, settle, die

2

A 2 B 3 C 5 D 1 E 4

Ex. 2

1 down **2** down **3** up **4** up **5** up **6** up
7 up **8** down

Ex. 3

2, 3

Ex. 4

1 breakdown **2** upset **3** feedback **4** clearout
5 takeaway **6** setback **7** lineup **8** turndown
9 outcome **10** outlook

Writing: report (Part 1) p.125

Aim:

- **to practise writing a report**

1 Students read the task and suggest which of the recommendations here are similar and might be grouped together in the report.

2 Remind students briefly of the layout of a report and the usual use of subheadings and bullet points. Then they look at the three plans and tell you which reflects the usual structure of a report.

3 Ask students to complete the advice individually by inserting *do* or *don't* and then to compare their ideas in pairs. Draw attention to the difference between 5 and 11; they should use all of the given information but avoid copying the same words.

4

1 Students skim the report ignoring the gaps and identify which of the plans in Exercise 2 the writer has followed. Then they work in pairs or individually to insert linking words.
2 Establish that the input information has been organised into groups, and ask the class to tell you what these groups are and to identify the sentence in the survey results which makes this clear. (*Suggestions were divided into recommendations for the day, and ideas that would have an effect in the long-term.*) Point out that this type of 'organising' sentence comes before the ideas themselves.

5

1 Students think how the suggestions could be fitted in. Remind them that they do not need to adopt all of the suggestions.

2, 3 Discuss these questions briefly with the whole class. Students then write the rest of the model report.

4 Students now swap their reports and read each other's. They should use the list of *dos* and *don'ts* in Exercise 3 to check the style, layout and content

6 This task can be set for homework.

ANSWERS

Ex. 2
Plan B

Ex. 3
1 do **2** do **3** don't **4** do **5** don't **6** do
7 do **8** do **9** don't **10** do **11** do

UNITS 6–10 Progress test p.127

ANSWERS

Ex. 1
1 C **2** B **3** C **4** A **5** A **6** D **7** B **8** C
9 A **10** C **11** D **12** B

Ex. 2
13 as **14** the **15** or **16** whose **17** not
18 out **19** so **20** At **21** but **22** since
23 where **24** getting **25** set **26** such
27 instead

Ex. 3
28 according **29** marketing **30** conferences
31 presentations **32** creative **33** communicator
34 developments **35** seasonal **36** genuinely
37 frustrating

Ex. 4
38 set **39** choice **40** draw **41** hold **42** loose

Ex. 5
43 wish (that) I'd/I had had
44 time (that) we had the car/time (that) the car was
45 on the point of sending
46 has become an obsession
47 was having children which (had) changed
48 widely thought to have been
49 the strength of the sun that/which
50 advised us to take

UNIT 11 Always on my mind

Grammar 1: modal verbs 2 p.130

Aim:

- **to review use of modals for talking about the past**

1

1 Write the word *memories* on the board and ask students if they can remember much of early childhood i.e. pre-school age. Ask what kinds of things they remember and why. Then ask them to skim the text and answer the gist questions.

2 Students work individually to identify the modal verbs and match the uses.

2 Students work in pairs to complete the sentences.

> **ANSWERS**
> 1 a) Gary b) Ian c) Helen d) Julie
> 2
> Possibility: I couldn't have remembered; it can't have happened; it might have been a dream; this may have been
> Logical deduction: must have been incredibly worried
> Obligation/necessity: I had to blow out …
> Advice: ought to have done …
> Permission: could blow out the candles
> Ability: I couldn't see my family; I can even remember thinking; she could quite clearly remember
>
> **Ex. 2**
> 1 might have/may have
> 2 can't have/couldn't have
> 3 could **4** should have **5** couldn't **6** had to

Vocabulary 1 p.131

Aim:

- **to review some idiomatic expressions with *take*, *mind* and *brain(s)***

1

1 Students match the sentences and then compare their answers in pairs.

2 Students discuss what they think the meanings of the expressions are. You could allow them to check in the dictionary before going through the answers.

2 Follow the same procedure as in Exercise 1. There may be expressions here that students already know, so you could ask them to scan the sentences and insert the expressions that they are familiar with first.

3 Students talk in pairs about the situations. Begin by choosing one of the expressions and telling a corresponding anecdote of your own and then asking students to identify which idiom it illustrates.

> **ANSWERS**
>
> **Ex. 1**
> **1** d **2** c **3** e **4** g **5** a **6** h **7** i **8** b
> **9** j **10** f
>
> **Ex. 2**
> **1** make up his mind **2** pick your brains
> **3** take your mind off **4** out of your mind
> **5** got it on the brain **6** read my mind
> **7** speak your mind **8** put your mind at rest
> **9** racking his brains

Exam focus

Paper 1 Reading: multiple choice (Part 3) p.132

Aim:

- **to complete a multiple-choice reading task**

1 Go over the exam information and procedure with students. Then ask them to skim read the text. Ask them one or two gist questions for this such as *Which paragraph gives anecdotes about two of the writer's friends?* or *Which paragraph talks about loss of memory in social situations?* Then students look at the multiple-choice questions, focusing on the stem of each one for the moment, and identify which paragraph they need to look in for the answers. In questions 3 and 5, of course, this is explicit and in some of the other questions, there are proper nouns which can easily be seen in the text. Then give students about ten minutes to read the text again, stopping at the appropriate place for each question to choose the correct answer. Finally, students compare answers in pairs and discuss any differences before checking as a whole-class activity.

2 As well as discussing memory in general, this is also a good opportunity to discuss how students remember words and expressions in English and how they can organise their vocabulary notes to make them memorable (organising words according to collocations and topics, using diagrams and colours).

3

1 Students match the verbs and nouns first and then check their answers against the text.
2 Use this question to check their understanding of expressions 2 and 4, and then check the meanings of the rest.

ANSWERS

Ex. 1
1 A **2** B **3** A **4** D **5** D **6** B **7** C

Ex. 3
1
to stop – in her tracks
to slip – my mind
to press – a key
to build up – familiarity with something
to swim – into our consciousness
to dial – a number
to tell – anecdotes
to let – someone off easily
2
a) to slip my mind
b) to build up familiarity with something

Use of English 1: multiple-choice cloze (Part 1) p.134

Aim:

- **to complete an exam-style multiple-choice cloze task**

1

1 Students talk in pairs about how easily they can concentrate, answering the questions here. At the end, ask some individual students to tell the class some of the things that help or hinder their concentration.
2 Students now skim read the text. After about 15 seconds ask them if it contains any of the ideas that they discussed together.

2 Students work individually to choose the most suitable words and then compare answers in pairs. When checking the answers, encourage students to record any useful collocations or expressions such as *face up to a problem* and any words which require a particular preposition such as *devoid of*.

Teaching tips and ideas

To extend work on a multiple-choice cloze, ask students to take one of the wrong answers for each question or a given number of them and rewrite sentences from the text using those words. This is particularly useful when the correct answer depends on grammar or prepositions.

ANSWERS

Ex. 2
1 B **2** D **3** A **4** D **5** C **6** B **7** B **8** D
9 A **10** C **11** B **12** C

Grammar 2: emphasis with inversion p.135

Aim:

- **to focus on negative inversion for emphasis and provide controlled practice**

1

1 Students may have some knowledge of this area of grammar but they are unlikely to be very familiar with it. Ask students to look at the example and tell you what has happened to the order of the words in the second sentence. Then point out that a negative word, if placed at the beginning of the sentence for emphasis, requires an inversion of the verb (as in questions). Then students look at sentence pairs 1 to 5 and indicate the negative words and inverted verbs in each emphatic sentence.
2 Complete the rule together as a class.

Watch Out! *no sooner/hardly*
Hardly and *no sooner* have the same meaning in this context. They both require negative inversion but a different link word. Ask students whether *than* or *when* is correct in each case, if necessary prompting them by pointing out that *No sooner* is a comparative structure.

2 Students work in pairs or individually to transform the sentences.

3 This can be done as a written transformation exercise, as above, or, as it is easier to convert the sentences back to the non-emphatic form, you could do it orally.

4 Students work in pairs or individually to transform the sentences.

ANSWERS

Ex. 1

1

1 a) Seldom have I come across such a strange story.
2 b) At no time must you leave your bag unattended.
3 b) Not until I went into the garden did I realise how hot it was.
4 b) Under no circumstances must you go back into the building after midnight.
5 a) Not only did I hate the book, but I hated the film of the book too!

2

When words and phrases like *not only, under no circumstances, at no time, not until, seldom* and *hardly* begin a sentence the verb and subject are inverted.

Watch Out!

a) than
b) when
1 *than*
2 *when*

Ex. 2

1 Hardly had I sat down to read the newspaper when the telephone rang.
2 No sooner had she stood up to speak than the fire alarm went off.
3 Not only does he forget people's names, he also finds it hard to remember place names.
4 Under no circumstances should you (ever) let anyone into your house unless you have seen their ID.
5 At no time did she (ever) doubt that he was telling the truth.
6 Only after I started to write the letter did I realise that I had lost their address.

Ex. 3

1 You rarely find a household without a computer these days.
2 She had only just/hardly started to have a shower when the postman knocked at the door.
3 I left for the airport and then remembered that my passport was still in my desk in the study.
4 He trusted her, and he never doubted her loyalty to him.
5 We have never seen such rapid progress in medical science at any time in recent history.

Ex. 4

1 sooner had Jane arrived than
2 no circumstances must mobile phones be
3 do you come across
4 before had I tried

▶ Photocopiable activity 11A *Emphatic inversion* p.174

Speaking 1: individual long turn (Part 2) p.136

Aim:

- **to practise carrying out an exam-style speaking task**

1
Students read the instructions and identify the two elements of the task. Then they complete the speaking task in pairs. If you wish to make this more exam-like, you could time each student exactly one minute, or ask students to work in groups of three, with one member of each group doing the timing and two carrying out the task. Then, of course, they should swap roles.

2 This question resembles the short question that the interlocutor will ask the other candidate after the long turn has been completed. Point this exam procedure out to the class and ask the question briefly to one or two individual students.

Vocabulary 2 Phrasal verbs with *think* p.136

Aim:

- **to revise or introduce some phrasal verbs and idiomatic expressions with *think***

1 Students complete the prepositions exercise in pairs. Go over the answers, checking the meaning of each phrasal verb. You could also point out which phrasal verbs are separable and which inseparable (*think out, think up* and *think through* are separable as we say, for example, *think out the proposal, think the proposal out* and *think it out*).

2 Ask students to skim read the text and then replace the highlighted phrases. They could check any phrases they are unsure of in a dictionary before you go through the answers.

3 Students briefly talk together about a situation which illustrates each of the idioms. Tell the class about a situation of your own, if possible.

ANSWERS

Ex. 1

1 about 2 of 3 out 4 over 5 through 6 up

Ex. 2

1 think straight 2 he thought the world of
3 think positively 4 think on his feet
5 thinking outside the box
6 thought better of it

Use of English 2: gapped sentences (Part 4) p.137

Aim:
- **to complete an exam-style gapped sentences task**

By now, students will be familiar with this type of exercise. It could be completed in class or set for homework.

ANSWERS
1 loss **2** mind **3** sense **4** attention **5** term

Listening: sentence completion (Part 2) p.138

Aim:
- **to complete an exam-style listening task**

1 Students talk in pairs to discuss the questions. If they have difficulty beginning, then brainstorm some answers to question 1.1 with the whole class and write a couple of answers on the board (e.g. jokes, someone making a silly mistake, etc.). Give the students a few minutes to answer the other three questions in pairs then discuss the four questions as a whole-class activity.

2 Students read the gapped sentences and as they have done in similar tasks, try to predict the answers. Ask which questions they feel are the most predictable (probably 3 and 4).

3 Then they listen to the recording twice to fill in the gaps, comparing answers in pairs after the first listening.

▶ Recording script p.103

4 Ask students if their predictions were correct and conduct a brief class discussion on why laughter is important in life and how individuals and groups can try and make more opportunities for laughter in their lives both at work and at play.

ANSWERS
Ex. 3
1 crying **2** catching **3** magazine article
4 offensive **5** humour **6** colds/(in)flu(enza)
7 World Laughter Day **8** mobile phones

Writing: article (Part 2) p.138

Aim:
- **to practise writing an article**

1

1 After a few minutes of discussion on the features of a good article, ask students to look back at the list of ideas on page 52 in Unit 4 and compare the ideas there with their own. If your teaching situation allows it, you could also prepare for this activity by asking students to scan some newspapers and magazines in the week before this lesson and find an article which they like. They then bring this to class to discuss with their partner and explain why they think it is a good article.
2 Students now read the article, keeping in mind the features that they have just discussed and again talk briefly about how many of these are demonstrated in this article.
3 This can be done in pairs as an extension of the discussion in 2 above.
4 Ask the class as a whole to choose the best title. You could also ask them to suggest any good alternative titles.

2 This can be answered as a whole-class activity

3

1 Students find the expressions in the article and then tell you the meanings, using the context as necessary.
2 Students work individually to find the corresponding expressions. After checking the answers, you could ask them to choose two or three that they think will be useful and write further example sentences.

4 Students read the two tasks on p.189 of the Coursebook and choose one.

5

1, 2 The first task is relatively similar to the article that the students have just read. If students are working on this, ask them to note down just one or two of the most important ideas from this article, plus at least one different one of their own. They talk together to think of specific people or examples of specific situations which support these ideas. They then follow the procedure in the book for paragraphing them and checking against the points listed in Exercise 2.
The second task gives students the chance to write about the topic of happiness from a slightly different angle. In this case, students will probably not be able to pool ideas in the same way, as they will of course know different people. They should work individually for about three or four minutes to choose a person and note down some ideas about their character and why they are happy. They then work in pairs or groups to explain their ideas to each other and answer any questions that other students may have. They could continue to work in pairs to paragraph

the article together and make any further suggestions arising from the list in Exercise 2.
Briefly check that students have noted all the key parts of each task. Then they form pairs or groups with others who have chosen the same article.

3, 4, 5, 6 Writing the articles could be done in class or for homework. When they have finished or in the next lesson, they should swap with a partner for feedback. Checking the article against the grammar checklist could be done either individually or, if students are happy about this, by their partner. Remind them of the piecemeal editing technique as suggested in earlier units. Allow students to rewrite their articles if they wish before handing them in to you for the final check.

ANSWERS

Ex. 1
3
a) 2 b) 2 c) 3 d) 1 e) 4

Ex. 2
2, 3, 4, 7, 8, 10

Ex. 3
1 a) optimistic not pessimistic; b) other people do better than you do; c) there are always opportunities to be found
2
1 conversely 2 crucial 3 an illusion
4 a positive outlook 5 people who look on the bright side 6 an enviable state of mind
7 go for it

▶ Photocopiable activity 11B *Eureka moments* p.175

Speaking 2 p.140

Aim:

- **to practise exchanging information and explaining reasons in spoken interaction**

1 Students try the two experiments in pairs. For the second, you will need to provide pieces of coloured card. Students read the two explanations on pp.189 and 190 of the Coursebook, then close their books and explain them to each other in their own words. At the end, ask them if they know of any other optical illusions.

2 Students discuss the puzzle in groups and see which group can arrive at the correct solution first.

UNIT 11 Review p.141

ANSWERS

Ex. 1
1 Researchers **2** beneficial **3** establishments
4 maximise **5** packages **6** commercially
7 dominant **8** mobility **9** recall
10 performance/performing
2
maximum – maximise
dominate – dominant
search – researchers
3
Q1 the verb 'have' after the gap indicates a plural noun, whereas 'research' is uncountable – so the answer must be the people who do it – researchers.
Advice: look carefully at the grammar of the sentence around the gap.

Q3 needs a plural noun because 'establishment' here is a concrete noun meaning an institution, so is countable. The rest of the sentence 'and private companies' helps you to see this.
Advice: Think about the context of the sentence around the gap.

Q8 The negative prefix is wrong here as it contradicts the idea of 'get stuck in' in the previous sentence which means *immobile*. So people 'regain mobility' if they are stuck.
Advice: Think about the meaning of the whole text and check the sentences before and after the one with the gap to check that your answer makes sense.

Ex. 2
1 think straight **2** thought it through
3 think the world of her **4** think up
5 think outside the box

UNIT 12 A matter of time

Reading 1: multiple choice (Part 1) p.142

Aim:

- **to complete an exam-style reading task**

1, **2** Students briefly discuss the questions and then feed back their ideas to the class. They read the first text to confirm their predictions.

3 By this stage students should know how to approach this task. If you think the students need reminding of the procedure, refer them back to the exam advice in the Exam focus for Reading Part 1 on page 20.

4 You could focus on either or both of the two questions. For the first question, ask students to give some examples of events which could influence a person's present and future, such as deciding who to marry.

ANSWERS

Ex. 3

1 D **2** A **3** D **4** B **5** C **6** A

Vocabulary: idiomatic language/ collocations p.144

Aim:

- **to focus on some lexical items from the reading text and to introduce some expressions and collocations with time, and provide controlled practice**

1

1 Students work individually to find the words and phrases and then check answers in pairs.

2 Students can do the exercise individually, or you could do it orally as a whole-class activity. Encourage students to record the words and expressions in their vocabulary notes with other possible collocations e.g. *give weight to the theory/evidence/view.*

2

1 Ask students to explain the phrase *before its time* as used in the example. You could also introduce the near-synonym *ahead of its time.*

2 Students now work individually to complete the sentences. Encourage them to guess the meaning if they are unsure. Then they compare in pairs. Ask them to suggest any phrases with similar meanings, with or without the word *time*, such as *fill in time* for *kill time* and *time and time again* for *time after time.*

3 Begin this activity with an example of your own for which the class should guess the corresponding phrases. Then students do the activity in pairs.

ANSWERS

Ex. 1

1

1 highly dubious 2 mass production
3 art icon 4 giving weight to 5 glimpse
6 scrupulously 7 hit-rate
8 hanging around 9 in due course
10 ups and downs 11 ethos

2

1 highly dubious 2 gives weight to
3 mass produced 4 glimpse
5 hanging around 6 ups and downs

Ex. 2

1

it existed before any other examples of the same thing

2

1 waste of time 2 at the time
3 for the time being 4 in no time
5 pressed for time 6 kill time 7 in time
8 running out of time 9 time after time

Grammar 1: passives 1 p.144

Aims:

- **to practise using passive constructions, especially passive infinitives**
- **to focus on contexts in which passive forms are preferred**

1

1 Write the title of the text on the board and ask the class for suggestions on what the text is about or ask them to speculate in pairs. Then students skim read the text to find out.

2 Students work individually to rewrite the phrases. Tell them that they do not need to use 'by' in these contexts. At the checking stage, watch out for mistakes in word order where the subject in the passive sentences is long, such as **It should not be ignored this opportunity to capture living history.*

3 This can be done as a whole-class activity. For some of the sentences, more than one option may be true; for example, for question 6, the reason for preferring the passive could arguably be either a) or b).

2 Go over the two examples with the class, pointing out the use of present and past infinitive. Students work individually to complete the transformations and then compare in pairs before checking as a whole-class activity.

Teaching tips and ideas

Before attempting any key word transformation question, students should try to identify which area of grammar it is testing. Encourage students to keep an ongoing, separate record of the answers to any key word transformations from the Coursebook, photocopiable tests or previous CAE papers that they were unable to answer, which they can look back at from time to time. Once they feel that they can confidently answer a transformation which tests that particular lexical item or grammatical area, they can cross it out.

3 You may wish to focus just on those questions which your students are most likely to respond to. Students are likely to have more to say if the questions can be personalised a little; for example, for question 2, you could ask them if they remember any stories that grandparents or older relatives have told them about the past.

ANSWERS

Ex. 1

2 and 3

1 in which older people from diverse ethnic backgrounds are encouraged to share and record their memories a)
2 might otherwise be lost to people b)
3 Age Exchange was set up in 1983 a)
4 the Reminiscence Centre was opened a)
5 many vital jobs have been assigned to volunteers b)
6 they will be asked b)
7 This opportunity to capture living history should not be ignored. c)
8 a rich vein of living history will have been closed for future generations c)

Ex. 2

1 was to have been finished
2 bad behaviour to be repeated
3 to be seen (by everyone) as
4 believed to have moved away
5 is said to have recovered
6 is nothing to be done (*is nothing anyone can do* is also possible, but does not use a passive infinitive as required)

Reading 2 p.146

Aims:

- **to introduce the topic of time travel**
- **to give practice in discussing and giving opinions on books**

1

1, 2 Students read one of the synopses and then work in groups of three to tell each other about what they have read. Encourage them to keep their books closed at this point so that they put it into their own words.

3 This is best done as a whole-class discussion. Students may have read *The Time Machine* as H.G. Wells is likely to have been translated into their language. In this case, you can simply ask their opinion on the story. You could also ask them if they know any other books or films in which time travel takes place (there are a number of such books in English especially in children's literature).

2 The topic of the text should lead on naturally from the discussion above. Students skim read the text to find the author's opinion.

Exam focus

Paper 3 Use of English: word formation (Part 3) p.146

Aim:

- **to complete an exam-style word-building task**

1 Go over the exam information and procedure. This can be done briefly as students will be familiar with this type of exercise by now. Then students work individually to complete the exercise. You may like to give them a time limit of ten minutes for this. Go through the answers with the whole class, pointing out other words which can be formed from these base words such as *speculate – speculation – speculative*.

2 Use the two questions to conduct a brief class discussion to round off the activity.

ANSWERS

Ex. 1

1 convincing **2** objections **3** accidentally
4 unappealing **5** novelist **6** speculation
7 relativity **8** existence **9** unwilling
10 ridiculous

Grammar 2: the future in the past p.147

Aim:

- **to present/review ways of expressing the future in the past and to provide controlled practice**

1 Write an example of a sentence containing future in the past on the board e.g. *I was going to post the letter yesterday but I forgot*. Establish that the sentence is talking about the past (*yesterday*) and the past continuous here is a way of talking about something which was in the future at that time. The students look at the three examples. Establish that for future in the past, future events which would normally be expressed with a present continuous are expressed with past continuous, and future events which would be expressed with *going to* future are expressed with *was/were going to*.
You may also wish to point out the frequent use of the stressed auxiliary verb for future in the past when talking about plans which were unfulfilled or changed, e.g. *We were meeting at 6 o'clock but Joe called and changed it*.
Students finally look at the sentences a) to c) and identify which one expresses an unfulfilled plan. Ask them to try saying the sentences with a stressed *was*.

2 Students complete the dialogues individually. Then ask pairs to read them out to the class to check the tenses. Point out that the auxiliary *was/were* could be stressed in questions 4 and 5.

3 This exercise focuses on future in the past forms for *will*, *be to do something* and *be about to do something*. Do the first question together and point out to students that future events normally expressed with *will* can be expressed with *would* for future in the past and that *be to* and *be about to* become *was/were to* and *was/were about to*. Then students complete the exercise, individually or in pairs. When going through the answers, point out that question 3 contains a future in the past form of the future perfect.

4 Students work individually to correct the verb forms. If they have difficulty, tell them in each case to decide which future form would be used normally (present continuous, *going to*, *will* or future continuous) and then see if the text uses the corresponding future in the past form. They compare in pairs before checking answers as a whole-class activity.

5 Students talk in pairs about each of the three situations. At the end, ask some individual students to tell the class about one of the stories that they heard.

ANSWERS

Ex. 1

a) 2 b) 1 c) 3

Ex. 2

1 was thinking of seeing **2** was going to call
3 was coming/was working
4 was just going to go **5** were going to go

Ex. 3

1 wouldn't take **2** would be
3 would have been **4** was about to
5 were to arrive

Ex. 4

1 was going to find
2 was going to
3 was scheduled
4 would still be looking for/would still have been looking
5 would be able to
6 was

▶ Photocopiable activity 12 *White elephants* pp.176 and 177

Listening: multiple matching (Part 4) p.148

Aim:
- **to complete an exam-style listening task**

1 Students discuss the questions in pairs. Members of the class who are keen on computer games will probably have plenty to say; if, however, you have a class who know little about them, you might prefer to make it a brief class discussion on just the first two questions.

2 Again, if your students do not know much about computer games, this is probably best done as a whole-class activity. You could encourage them to think about what features they would need or wish to see as an inexperienced player.

3 Students will be familiar with this kind of task by now. Remind them to read the questions carefully first and concentrate on 1 to 5 on the first listening and 6 to 10 on the second. Encourage them to think about other ways in which these ideas could be expressed, e.g. how could the phrase *easy for beginners* be paraphrased? Remind them too that they may hear more specific information in the listening text than in the questions, e.g. the setting and characters could be named. Play the recording twice and students do the task. They compare in pairs before checking as a whole-class activity.

▶ Recording script p.103

ANSWERS

Ex. 3
1 E 2 A 3 H 4 G 5 B 6 F 7 E 8 B
9 D 10 G

Speaking: two-way conversation (Part 3) p.148

Aims:
- **to carry out two exam-style speaking tasks (Paper 5, Parts 3 and 4)**
- **to revise language for inviting opinions and encouraging your partner to speak**

1 Students look at the pieces of advice and decide which one is misleading. Then ask the class how they would change this to make it good advice (e.g. *Only make a decision when you have listened to your partner's opinions.*).

2 Students brainstorm together or as a class some phrases which could be used for inviting their partner to give his/her opinion.

3 Students talk for approximately four minutes to discuss opinions and make a decision. Remind them that they will need to use third or mixed conditionals for this task (e.g. *If cars hadn't been invented, we would still be travelling on horseback.*).

4 These questions resemble the sorts of questions that the interlocutor may ask in the final part of the speaking exam. Students could continue to work in pairs or you could act as interlocutor and ask the questions to different pairs as a whole-class activity. Students may not know any definite answers for question 3; if so, encourage them to make a general statement about how it can be difficult to know how important a new invention will be or what effect it will have. Computers are a famous example of this as the idea of everyone having a home computer was dismissed in the early days. Other inventions which were dismissed or ridiculed when they were first proposed were submarines and postage stamps.

5 If Exercise 4 was done in pairs, students should now tell the class some of their ideas.

Use of English: open cloze (Part 2) p.150

Aim:
- **to complete an exam-style open cloze task**

1

1, 2 Write the topic *science fiction* on the board and ask students to give some examples of books or films which come under this category. (They have already read the synopsis of *The Time Machine*.) Then ask if they enjoy this type of book or film and what value it can be to us. It is usually argued that science fiction can warn us what the outcome of our present situation might be or about the dangers of certain kinds of technological advance which can go out of our control (*Frankenstein* is a famous example of this).

3 If students do not know *Star Trek*, ask them to speculate on what the programme is about from the photograph before asking them how accurate it might be.

2 Students now skim the text to find out the writer's opinion.

3 Students work individually to complete the text and then compare answers in pairs. When going through the answers, encourage them to note any useful collocations or expressions such as *catch up with*.

4 Students think of other examples of science fiction and compare them with technological advances that have actually happened as in the text.

ANSWERS

Ex. 3

1 since **2** as **3** According **4** up **5** in **6** and **7** Such **8** no **9** before **10** which **11** would/might/could **12** a **13** take **14** one **15** if

Writing: essay (Part 2) p.151

Aim:

- **to give practice in writing an argumentative essay**

1, **2** Go over the information about the possible structures and conventions for argumentative essays with students. Then students complete the table comparing the features of essays, articles and reports. Check the answers and ask students if they can suggest other typical features of each genre.

3 After students have read the task, ask if they agree or disagree and why. Write their ideas on the board, divided into reasons why the past should be preserved and reasons why there is no point. Develop these ideas by eliciting concrete examples of buildings and stories which would be worth preserving and possibly also ones which would not. Students then write the essay, either in class as a timed exercise or for homework.

4

1 Students read the example answer and pick out which of the three possible ways of organising it the writer has used. Point out the use of rhetorical questions to reinforce an opinion or challenge a statement and ask students to find further examples in paragraphs 2 and 3.

2 Complete the outline of the essay together as a class.

3, 4, 5 Students now focus on the language of the essay. They underline the link words and check they understand the function of each. Then ask them to suggest alternative link words that could be used.

5 Students work in pairs to develop the paragraphs further. If they have difficulties, suggest that they think of a concrete example of the type of person who looks back too much or something specific about the past that is useful to understand.

6 This can be set for homework. If the students are confident about this type of writing task, you could give it to them without doing any class preparation. Otherwise, you could plan it in class beforehand following the steps in Exercise 4.2 above.

ANSWERS

Ex. 2

essay	article	report
language		
uses semi-formal language	uses colourful language	uses objective language, often with the passive
purpose		
persuades through discussion	entertains	makes recommendations based on facts
organisation		
presents a clear argument in linked paragraphs	*uses paragraphs for effect*	can use bullet points
target reader		
known reader	general reader	known reader

Ex. 4

1

1 Plan A
2 semi-formal
3 to introduce the argument

2

– there is a strong argument for looking forwards, not backwards
– we can't change the past, it is where we came from

3

the first idea: on the one hand
a contrasting idea: on the other hand
an opinion: it seems to me
a reason: so/for this reason

4 surely/obviously: they show the writer's own opinion
too: it indicates that we understand our background at the moment, but that future generations won't

5 general statement: some people say
a rhetorical question: But is this really true?/And surely we have a responsibility to future generations, so that they too can understand their background? sentence in the conclusion that links back to the introduction: For this reason, I feel that there is a lot of point in preserving old buildings and stories from the past, and that the statement is wrong.

UNIT 12 Review p.153

ANSWERS

Ex. 1

1 set **2** time **3** reached **4** sense **5** meant

Ex. 2

1 had **2** could have predicted
3 made **4** wanted to **5** happen

Ex. 3

1 held up for
2 have been reported to
3 original intention was to
4 is running out
5 are still thought to exist/are thought still to exist/are thought to still exist
6 was about to be a
7 (high/about) time (that) John got/had his
8 was scheduled to land

UNIT 13 A way with words

Reading: multiple matching (Part 4) p.154

Aim:

- **to complete an exam-style multiple-matching task**

1 Introduce the topic of celebrities by using the lead-in questions. If one is available, you could bring in a copy of a magazine which deals with celebrity lifestyle or gossip, such as *Hello!*, to illustrate which kind of stories they contain and why people might want to read them.
Then move on the topic of autobiographies, again using the lead-in questions in the book. If any student has read a celebrity autobiography, ask them what they thought of it.

2 You will need to pre-teach the term *ghost writer.* Students scan the text first for the proper names and pick out the names of the publishers, ghost writers and celebrities. Then they skim the whole text to gain an overall idea of the content. You could set a gist question for this such as *How has the market for celebrity autobiographies changed?*

3 Give students a few moments to look at the text again and identify the sections.

4 Remind students briefly of the procedure for this type of task as outlined in Unit 5 page 56 and then give them approximately ten minutes to complete it.

5 Students work individually to match the phrases. They then check their answers against the context in the text before checking as a whole-class activity.

ANSWERS

Ex. 4
1 E **2** B **3** A **4** D **5** AD **6** DA **7** BC
8 CB **9** A **10** D **11** B **12** C **13** E **14** C
15 E

Ex. 5
1 e) **2** i) **3** g) **4** c) **5** a) **6** b) **7** j) **8** k)
9 l) **10** f) **11** h) **12** d)

Vocabulary 1: adverbials expressing attitude p.156

Aim:

- **to focus on adverbs which express the speaker's attitude**

1 Students match the adverbs and meanings and then check together in pairs. After establishing that *clearly* and *obviously* have similar meanings, ask the class to suggest any other ways of expressing these feelings and attitudes. These could be other adverbs or adverbial phrases such as *to be honest*.

2 Students complete the exercise in pairs and then check in pairs.

3 Give students one or two minutes to think of four ideas for their statement. Then they work in pairs to discuss them. To encourage the use of the adverbs, you could put a list on the board and tell students that they cannot finish their discussion until they have used at least three of them.

ANSWERS

Ex. 1
1
1 Frankly f) 2 Actually g) 3 basically e)
4 apparently a) 5 personally d) 6 clearly b)
7 obviously c)
2 clearly/obviously

Ex. 2
1 obviously **2** basically **3** actually **4** frankly
5 apparently **6** clearly **7** Obviously **8** Basically

Grammar 1: participle clauses p.156

Aim:

- **to focus on the use of participle clauses to replace relative pronoun + verb**

1 Go over the initial explanation with students and then ask them to look at examples 1 and 2 in Exercise 2. Complete the rules about participle clauses together.

2 Students rewrite the sentences and then check answers in pairs. If they are confident about this grammar, you could do the second part of the exercise orally.

3 Ask the class to look at the example sentences and again complete the rules together.

4 Students rewrite the sentences and then check their answers in pairs

5

1 Give students 15 seconds or so to skim the text and then ask them to summarise the two main arguments that it refers to.
2 Students rewrite the passages in the text using participle clauses. You may need to warn them that this may involve shortening some of the clauses.
3 Students now discuss the content of the passage as a class. Ask them to think back to their first experience of learning another language and see if they can remember any experiences which might support one of the theories.

ANSWERS

Ex. 1

1

1 a) and 2 a) use a relative clause
2 Pair 1 is active, pair 2 is passive
active ... present participle
passive ... past participle

Ex. 2

1 The witnesses saw a fire burning in the distance.
2 Not all the people registered for the conference actually attended.
3 The piece of wood holding the window open had fallen out allowing the burglar to enter the house.
4 Your priority when making a career move should be the salary.
5 We obtained a copy of the government report published last week.
6 Anyone wanting to join should register on the website.
7 Working at a desk piled high with papers, his working conditions were clearly poor.

Ex. 3

1 a) Since (reason)
2 a) Once (time)
3 a) so (result)

The words are replaced in b) by participle clauses. Words such as *so*, *once* and *since* can be replaced by a participle clause.
When the sentence is in the past, the participle clause can be formed by *having* + past participle.

Ex. 4

1 Having arrived very late, we decided to get a taxi from the airport.
2 Having seen what the food was like in the hotel, I went to a restaurant to eat.
3 Having announced their proposals, the management expected the workforce to support their new pay structure.
4 Having looked forward to the party for weeks, I was upset when it was cancelled.
5 Having finished his lunch, he rushed out of the house to catch the train.
6 Having ordered a steak, I was annoyed when the waiter brought me fish.

Ex. 5

1 the idea that we all are born with language
2
(1) having been born/being born with an innate knowledge of grammar
(2) having started to speak
(3) showing
(4) copying models
(5) learning the rules

Speaking: discussion (Parts 3 and 4) p.158

Aim:

- **to carry out an exam-style speaking task (Paper 5, Part 3)**

1 Students talk in pairs for about four minutes about the photos and make their decision. Ask one or two pairs to tell the class their decision and why.

2 For these general questions, students could either discuss in pairs again or you could take the role of interlocutor and nominate individual students to answer them. At the end of the activity, give a short feedback session on any vocabulary or grammar errors.

Exam focus

Paper 3 Use of English: multiple-choice cloze (Part 1) p.158

Aim:

- **to complete an exam-style multiple-choice cloze exercise**

Students will now be familiar with this type of exercise. Go over the exam information and suggested procedure and ask if they have any more advice to add.

1 Students then complete the task. To make it more exam-like, give a time limit of ten minutes. When going through the answers, encourage them to note any useful collocations or phrases with prepositions in their vocabulary notes (e.g. *come up with an unusual way to do sth, tolerant of*).

2 Briefly discuss the questions with the class. Ask if they already take part in any activities which help them to relax or concentrate better, such as yoga.

ANSWERS

Ex. 1

1 A 2 B 3 B 4 C 5 A 6 D 7 A 8 B
9 C 10 D 11 B 12 B

Vocabulary 2: communication – idioms p.160

Aim:

- **to introduce idioms on the theme of talking and communication, and phrases based around the verb *say***

1 Give students one or two minutes to read the sentences and decide on the meanings of the idioms. Then they compare their ideas in pairs before checking with the dictionary.

2 Students work in pairs to describe and guess example situations for the idioms. Do one example with the whole class first.

3 Match the phrases and functions as a whole-class activity. Then students write dialogues in pairs. Ask one or two pairs to read them to the class with natural stress and intonation.

4 This could again be done in pairs or you could make it more like part 4 of the CAE speaking paper by doing it in open pairs. Some of the expressions are easier to use in an exam context than others. *I couldn't say* or *Who can say?* are probably unsuitable to use on their own as they will block the discussion. To counteract this, introduce some ways in which they can be modified as in *I couldn't say for sure but* … or *Who can say? It's so difficult to tell but* …

ANSWERS

Ex. 1

1 used to say you could have described something or criticised someone more severely than you have
2 polite friendly conversation about unimportant subjects
3 talk angrily – tell someone off
4 reach the most important part of what they want to say
5 understand
6 talk about work
7 misunderstood something
8 get a chance to speak
9 two people don't understand each other because they are talking about different things but haven't realised it
10 talk to someone as if they are stupid
11 says exactly what she thinks
12 use a metaphor to explain

Ex. 3

1 b) 2 d) 3 e) 4 a) 5 c) 6 f)

Grammar 2: passives 2 p.161

Aims:

- **to focus on contexts in which passive forms are commonly used**
- **to practise forming passive sentences**

1 Students work individually to match the sources and reasons. You may need to pre-teach the term *press release*. Some of the reasons overlap so that more than one possible answer might be a possible match for the reasons; however, encourage students to choose what they think is the main reason.

2 This exercise focuses further on reasons for using the passive in written English. Students read the title and headline of the text and tell you what *round the clock* means. Then they skim read the text to find out the reasons why this scheme has been proposed. Finally, they match the uses of the passive and compare their answers in pairs.

3 Students transform the sentences to the passive and then compare in pairs. When going through the answers, discuss why some of them are not possible or unlikely in the passive. To round off the topic ask students to find further examples of the passive and again discuss as a class why it is preferred in that context. Ask the class for their opinions of the proposal. This could lead into a discussion of other changes they would like to see or would have liked, as regards the start and finish times of lessons at school.

4 Students read the situation and brainstorm ideas in pairs, then compose a short paragraph. Give them some prompts for using the passive in this (e.g. *It is hoped that … It is believed that … We recommend that … should be …*).

ANSWERS

Ex. 1

Extract	Sources	Reason for the passive
Man bitten by dog	newspaper headline	object or event more important than subject
The house was …	guidebook	information more important than agent
The recom-mendation …	report	needs to sound objective and impersonal
The man was charged …	criminal record	charge more important than the person making it
It is hoped…	press release	no agent – object more important than subject

Ex. 2

1 d) **2** b) **3** c) **4** a)

Ex. 3

1 All your questions will be answered very soon.
2 It is said that …
3 ✗ (subject is Roman history)
4 It is believed that
5 I was sent a cheque for overpaid tax by the Inland Revenue.
6 ✗ (it's an order)
7 More than twenty students are going to be sent abroad to study by the college.
8 ✗ (no agent mentioned)

Listening: multiple choice (Part 1) p.162

Aim:

- **to complete an exam-style listening task (Paper 4, Part 1)**

Exam information

In Paper 4 (listening), students write their answers on the question sheet while they listen, and then are given five minutes to transfer the answers to an answer sheet at the end. It is a good idea to give students two or three practice runs in this process before the exam. Emphasise that they need to check after each section that they have not missed out any answers (if they have, the question numbers on the answer sheet will not correspond to their intended answers), that their spelling is correct and, in the case of Part 2, that what they have written fits exactly into the gap.

1 Ask students to talk in pairs and list at least four qualities of a good teacher. If you prefer, give them some examples and ask them to rank them in order of importance (e.g. *makes the student work hard, listens to problems, knows their subject well, good at discipline, has a sense of humour*). They then compare and discuss their ranking in pairs. The question as to whether education should be work-focused or not is a complex one. You could start by asking students to think about the education system in their own country and how work-focused it is and then whether they think it should be changed.

2 As students have practised this type of listening exercise before, you might decide to do it under exam conditions at this stage in the course, playing each extract twice through and only checking the answers at the end. Allow students to look at the suggested procedure in Unit 4 page 45 before beginning.

3 Ask the class if they agree with Dr Ashby and what effect they think new forms of media such as text messaging have on children's ability to communicate in speech or writing.

4 Move the discussion on to other types of new technology and the skills that they develop, or the skills that have consequently been lost.

5 Play the final extract again, stopping just after each word or expression and ask the class how they are used. You could ask them to write example sentences for some of the items, especially *get to grips with*.

▶ Recording script p.104

ANSWERS

Ex. 2
1 A 2 C 3 B 4 A 5 C 6 B

Ex. 5
individualistic – negative
self-centred – negative
hype – negative
get to grips with – positive

Vocabulary 3: similes (*like/as*) p.163

Aim

- **to introduce similes with *like* and *as* and provide contextualised practice**

1 Students predict the meanings of the similes from the contexts and then compare ideas in pairs before checking with the dictionary. You could ask students if they have any similar expressions in their own languages; *sleep like a log* in particular is likely to have a corresponding simile in other languages.

2 Students match the similes and then write sentences. You could extend the activity by asking one or two of them to read out the sentences to the class, blanking out the simile so that the other students can guess it. You could also again compare these with any corresponding similes in the students' own languages.

3 These can be completed in pairs, or done orally as a whole-class activity.

ANSWERS

Ex. 1
1 angry 2 slept very well 3 out of place
4 makes him angry quickly 5 insensitive, reckless

Ex. 2
1 b) 2 c) 3 a) 4 e) 5 d)

Ex. 3
1 like a fish out of water
2 as cool as a cucumber
3 as white as a sheet
4 like a log
5 like a bull in a china shop

▶ Photocopiable activity 13 *Gapped sentences quiz* pp.178 and 179

Writing: proposal (Part 2) p.163

Aim:

- **to practise writing a proposal**

1 Ask the class to read the list and identify the statements that are not true. You could also ask what other genre these statements are true for (report).

2

1 Students read the task carefully. Check their understanding by asking them to summarise the situation (unsatisfactory results in language exams) and what the proposal must do (suggest what should be done and why).
2 The class identify the best organisation.
3 Students work in pairs to brainstorm possible ideas for reasons and recommendations. After about five minutes, conduct a general class feedback and put the ideas on the board.

3 Students read the proposal and identify which of the two organisations in Exercise 2.2 is used. They use this to fill in the headings. Then ask the class to summarise the recommendations and reasons and compare them with the students' ideas on the board.

4

1, 2, 3 These questions are best done together as a class. Point out that the participle clauses in this case are also passive structures. You could also point out the formal second conditional structure at the end (*if they were to be implemented*).
4 Students work individually to proofread the proposal for spelling mistakes and then compare answers in pairs.

5 Students choose one of the tasks and then brainstorm ideas together. Suggest that they brainstorm under three headings, *issues, recommendations* and *reasons.* They then match the items under the three headings, possibly by drawing lines between them, to ensure that there is a clear correspondence, e.g. between issues and consequent recommendations. The actual writing can be done as a timed exercise in class or for homework.

ANSWERS

Ex. 1
Not true: It is written in an informal style;
It uses a range of interesting vocabulary.

Ex. 2
2 A

Ex. 3
1 Main issues
2 Recommendations with reasons

Ex. 4
1
The second problem is connected to this
DVDs should be made available
the language being studied
A monthly prize could be awarded
would need to be fully supported

These are in 'Recommendations and conclusions'. This needs to be more objective to carry more weight and to show that they are based on evidence and not just personal opinion.
2
The survey conducted among students
This would provide motivation, giving students a real reason to listen
3
~~don't get the chance to~~ rarely have the opportunity to
~~it could be lots of fun~~ it could be very enjoyable
4
~~conected~~ connected
~~dificult~~ difficult
~~unatural~~ unnatural
~~borow~~ borrow
~~progres~~ progress
~~esay~~ essay
~~recomendations~~ recommendations
~~suported~~ supported

UNIT 13 Review p.165

ANSWERS

Ex. 1
1 full **2** such **3** making **4** own
5 how/why **6** as/so **7** His **8** up **9** what
10 to **11** again **12** either **13** by **14** other
15 itself

Ex. 2
1 being translated into a variety
2 is (now) (being) more widely spoken
3 is thought to have developed
4 ought to/should be taught to
5 was asked to give an
6 notice is taken of elderly people

Ex. 3
1 like a ~~horse~~ bull in a china shop
2 as red as a ~~rose~~ beetroot
3 as strong as an ~~elephant~~ ox
4 like a fish out of ~~sea~~ water
5 it came to me in a ~~bang~~ flash
6 I don't know what to ~~tell~~ say to you/tell ~~to~~ you

UNIT 14 It's how you tell it

Reading 1: multiple choice (Part 1) p.166

Aim:
- **to complete an exam-style reading task**

1 Ask students to write down the names of any stories that they have read or heard recently whether on TV, radio, in newspapers, magazines or books. Ask them also to write down the name of a story they have read in their lifetime which they particularly enjoyed. Then they talk in pairs, saying what stories they have heard or read and then tell each other the outline of their favourite story.

2 Ask students to write down what makes a good story and then compare in pairs. Each pair can then feed back what makes a good story. Write these ideas on the board and encourage a whole-class discussion on good story writing.

3 Students now carry out the reading task. At this stage in the course, you may choose to do this as a timed exercise. Remind students of the suggested procedure, or allow them to look again at Unit 2 page 20 and then give them just over 15 minutes to complete the whole task.

4 Students discuss the question in pairs. If your students do not read fiction very much, you may choose just to focus on the first question.

ANSWERS

Ex. 3
1 A **2** B **3** B **4** D **5** A **6** C

Use of English: open cloze (Part 2) p.168

Aims:
- **to complete an exam-style open-cloze task**
- **to read a short poem and discuss reactions**
- **to experiment creatively with language**

1 Discuss the questions briefly with the class as a whole and then ask students if they have ever been to a poetry reading.

2 Students read the title and then give some examples of what they think count as 'popular culture'. Then they gist read the first part of the text to find out how popular culture began.

3 Students complete the task individually and then compare answers in pairs. If you wish to make it more like an exam exercise, you could give a ten-minute time limit.

4

1 Ask these questions to the whole class to elicit a summary of the content of the text. You could also ask students to give some further examples of poems originally written to be read aloud. (There are many other classical examples in many languages.)

2, 3 Students talk in pairs about their liking for or dislike of poetry and then decide which of the features in question 3 are most important in a poem.

5

1 Students work in pairs to put the lines together to create their poem. If you wish, you could allow them to add punctuation (there is none in the original). To encourage imaginative titles, you could tell students that their title must not contain any of the words in the poem.

2 In a large class, you may prefer students to work in groups, and read their poems to each other.

3 Students now hear the original poem in the recording to compare with their own. As the poem contains none of the features listed in Exercise 4.3, it may provoke some discussion about whether it is really poetry. You could ask students what, if anything, it would lose if it were written out as a piece of prose or a note.

▶ Recording script p.105

ANSWERS

Ex. 3
1 such **2** may/could/might **3** there **4** would
5 who **6** on/down **7** into/within **8** as
9 of **10** a **11** that/which
12 (Al)Though/While(st) **13** yet **14** like **15** its

Ex. 4
1
1 word of mouth
2 poetry changed
3 not very popular – occasional performances

Ex. 5

This is just to say

I have eaten
The plums
That were in
The icebox

And which
You were probably
Saving
For breakfast

Forgive me
They were delicious
So sweet
And so cold

(William Carlos Williams)

Vocabulary 1: books and stories p.169

Aim:

- **to introduce some lexical items on the topic of books and types of reading**

1 Students work individually to complete the sentences. You could allow them to check with the dictionary before going through the answers.

2 Draw students' attention to the verb *flip through* in the previous exercise and what kind of reading this means. Then students complete the sentences, using a dictionary if necessary. When going through the answers, check the meaning of the expression *from cover to cover* in question 3. You might also point out that *wade* and *dip* are both verbs for moving through water and that we sometimes use other water expressions to describe words, as in *a flood of words.*

ANSWERS

Ex. 1
1 volumes **2** copies **3** best-seller **4** edition
5 whodunit **6** thriller **7** paperback **8** blurb

Ex. 2
1 browsing **2** wading through
3 dipping into **4** skimming

Exam focus

Paper 5 Speaking (Parts 1–4) p.170

Aim:

- **to give the opportunity to carry out a complete mock speaking exam**

Students should already have a fairly good idea of what the CAE speaking exam involves but it will still be useful to bring it all together by going over the exam information and procedure as a whole. At the end of the procedure, ask students if they have any other favourite tips for succeeding in the speaking exam.

Put students in groups of four, or if this does not work out exactly, have one or two groups of five with two students as assessors. They decide who is to be the candidate and who the interlocutor and assessor(s) and work through the complete exam task. At the end, they discuss as a group what could be improved. If students have not had enough, you could run through the exam again with a different pair of students acting as candidates, or you may have access to CAE past papers so that a different exam set can be used.

Listening: multiple choice (Part 1) p.170

Aim:

- **to complete an exam-style listening task**

1 Students read the multiple-choice questions and predict as much as they can of the content of the listening texts. They note their answers to the questions and then compare in pairs.

2 At this stage, the task is probably best done under exam conditions.

▶ Recording script p.105

3 Use one or more of these questions to conduct a brief class discussion.

ANSWERS

Ex. 2
1 B **2** A **3** C **4** B **5** A **6** B

Vocabulary 2: synonyms p.171

Aim:

- **to extend vocabulary by focusing on synonyms and paraphrases for common expressions**

1 Students work in pairs to think of synonyms to replace the underlined words. If they find this difficult, put a list of jumbled synonyms on the board or OHT and ask them to select from these, using dictionaries to help them if necessary. The list could include the following: *beg with, plead with, frightened, on edge, realise, understand, be aware, consequently, because of this, wonder, agonise, ponder, have* *an idea, cannot imagine, startle, panic, jump out of her skin.* At the end, students compare their version with another pair's.

2 Students now continue the narrative by writing the next paragraph. If necessary prompt them with questions such as *Who was ringing? Why? Did she stay in the hotel? Did she contact Carlo again?* Pairs then read their paragraphs to each other. You could ask each pair to suggest synonyms for two of the lexical items in the paragraph they hear.

3 Students now listen to an example paragraph. Allow them to listen for the gist the first time. Then play the recording again and ask them to write down any interesting vocabulary that they hear. Allow them to use dictionaries to check any further contexts or collocations.

▶ Recording script p.105

ANSWERS

Ex. 1

Ex. 1 sample answers

1 Janet begged/pleaded with Carlo to take her back to the hotel. She felt frightened/afraid/terrified now that she knew that the killer was still free. It was all because she had information that could convict him, and she now realised/understood that he was aware/realised – so because of this/consequently her life was in danger.
Once she arrived in her hotel room she sat down on the bed, wondering/agonising over/pondering what to do next. She didn't have a clue/have any idea. She panicked/jumped/jumped out of her skin when the telephone rang suddenly. She picked up the receiver, her hand shaking. The voice at the other end was strangely familiar.

Ex. 3

pounding; from head to foot; burst in

Writing: the set book (Part 2 question 5) p.172

Aims:

- **to prepare for questions that may be asked about the set book**
- **to practise writing an essay on the set book**
- **to think about aspects of books that could be helpful in answering other writing tasks**

1 If students are not studying the set book and have not read the same books, ask students to think about a book they have read and note down some answers to the questions in Exercise 1.2. Then they talk in pairs about the books that they have chosen.

Teaching tips and ideas

Characters or events from books and films can often provide good examples to support students' ideas and opinions, not just in writing about the set book, but also in other types of text such as articles and argumentative essays.

2 Students read the two tasks and the example answer. They then work in pairs to answer the questions in 3.

3 The writing could either be done in class as a timed exercise or set for homework with the proofreading and checking taking place in the subsequent lesson.

ANSWERS

Ex. 2

3

1 a whodunit/thriller
2 semi-formal
3 businessman, dilemma, struggle, reservation

4

a) a real page turner
b) the story is full of twists
c) stumbles across
d) a quick read

Grammar: mistakes to avoid p.173

Aims:

- **to highlight common mistakes and focus on words students commonly misspell**
- **to revise the use of punctuation marks and provide controlled practice**

1 Go over the spelling rules with students. For the *i* before *e* except after *c* rule, you may need to tell students that it only applies when the combination is pronounced /iː/ as they will probably think of a number of common words where it does not apply such as *their* and *eight*. Then students do the spelling exercise and compare their answers in pairs. When going through the answers, point out the usefulness of looking at the prefixes and suffixes as a spelling aid; for example *useful* = *use* plus the adjective suffix *ful*.

2 In addition to the pairs exercise, you could also take some words which have been frequently misspelt in the students' written work and give them to students as a spelling test.

3 After students have discussed the punctuation issues in pairs, conduct a brief feedback session to see if they have any punctuation errors in common. One tendency which students often have is to use too many commas and not enough full stops.

Watch Out! *apostrophe*
This section revises the basic two uses of apostrophes. Students often fail to distinguish between singular and plural in their use of apostrophes or may put them on simple plurals by mistake. You may like to tell students that English people sometimes get confused with them as well.

4 Students complete the exercise individually and then compare answers.

5

1 If students are not sure which area of punctuation to focus on, you could ask them instead to write three sentences, one with a deliberate punctuation mistake and ask a partner to spot the sentence with the mistake.
2 Students could do this exercise in pairs in class or it could be set for homework.

ANSWERS

Ex. 1

1 recommendation
2 believe
3 incredibly/disappointing
4 because/movement
5 argument
6 their/choice
7 practise/supple
8 comfortable/afford/ourselves
9 What/watching
10 benefit/languages
11 psychology/useful

Watch Out!

1 a, 2 b, 3 b
It's = It is
You're = You are
Video's = possessive
Videos = plural

Ex. 4

1 I do like going to the cinema; however, I really didn't enjoy the last film I saw.
2 My favourite English meal is fish and chips, although I also like roast beef.
3 I'm meeting my friends at the theatre because they're probably going to arrive quite late.
4 My five-year-old nephew loves reading – it's really good that he does.
5 It was your idea to go out last night so it's not my fault that you are tired this morning!
6 People say that young people don't read much nowadays, so how often do you read a book?
7 I find it very strange the way English people eat potatoes with every meal!
8 I couldn't answer the last question in the test so I asked the teacher afterwards.
9 The reviews of the film were really good, so I was very disappointed that I couldn't go.
10 Let's go to the show tonight instead of Saturday, because I'm really looking forward to it and I can't wait!

Ex. 5

1

This book grabbed me from the first page; whenever I had to put it down, I couldn't stop thinking about it. It's not unlike a Sherlock Holmes novel, but set in New York and not England. The story's full of twists and turns; these kept me guessing the whole time. The author's attention to detail brings the city vividly to life. Overall it was a thrilling, exciting read and I'm sure you're going to love it!

▶ Photocopiable activity 14 *Spelling* pp.180 and 181

Writing: Paper 2 overview p. 174

Aims:

- **to provide a review of possible question types in CAE Paper 2**
- **to focus on how written work will be assessed in the exam**

1 Go through the exam information with students, possibly reminding them of when they practised each genre.

2 Ask students to choose what for them are the easiest and most difficult question types and then compare their ideas in pairs or groups.

3 Students read the task. Ask them to underline the sentence in the announcement which gives the instruction.

4 Students read the answer and the band descriptors. Ask them which parts of the descriptors refer to language and which to the content of the task. Then they decide on the correct band in pairs. Discuss their decision as a whole class and ask them to justify it.

5 Students work in pairs to improve the example answer. After about ten minutes, they join up with another pair and read their improved answer to each other.

6

1 Students read the three tasks and underline the essential parts of the task in each one. Go over the advice with the class.
2 Allow students a few seconds to choose one of the tasks and possibly to begin making notes. Then they talk to a partner about their choice and what they have decided to write about.
3 Students complete the writing task they chose under timed conditions. This could be set for homework, although it is then often tempting for students not to observe the time limit.

7 Following this or in a subsequent lesson, students swap and assess each other's answers. Ask them to read the answer carefully at least twice, first to assess the content and secondly to assess the language, possibly using a grammar checklist.

8 In response to their partner's feedback, students can choose either to hand in their answer or rewrite it for the following lesson.

Reading 2: planning to take an exam? p.176

Aims:

- **to focus on preparation in the weeks running up to the exam**
- **to practise identifying the overall topic of a paragraph**

1

1 Students look at the headings and discuss with a partner the kind of advice that they would give for each one. At the end, ask each pair to tell the class one of their pieces of advice. You might want to tell them heading d) should focus on things to do outside the classroom and heading f) is about not becoming tired too quickly.
2 Students now match the headings and complete the advice. When going over the answers, ask if they have anything to add to any of the sections.

2 If students find it difficult to pick out one piece of advice, as an alternative you could ask them which piece of advice would help most for each paper in the exam.

ANSWERS

Ex. 1

2

A Being aware of your particular issues
1 Do 2 Don't
B Looking after your body
3 Don't 4 Do
C Getting into the swing of it
5 Don't 6 Do
D Making the most of past papers
7 Do 8 Don't
E Keeping yourself going
9 Don't 10 Do
F Getting things into your head
11 Do 12 Don't 13 Do

CAE quiz p.177

Aim:

- **to fill in any gaps in the students' knowledge of the CAE exam and how it is assessed**

Give students about five minutes to complete the quiz individually. Then go through the answers with the whole class.

ANSWERS

1 Five: Reading, Writing, Use of English, Listening and Speaking
2 Reading: 1 hour 15 mins, Writing: 1 hour 30 minutes, Use of English: 1 hour, Listening: 45 minutes, Speaking: 15 minutes.
3 20% of the total
4 No, you do not get separate marks for each paper, your result is based on an aggregate.
5 Yes, but the examiner will not see them. Only the answers you enter on the marksheet will count.
6 No
7 All marksheets (Papers 1, 3, 4) should be completed in pencil. All answers to Paper 2 (Writing) should be in pen.
8 You can choose.
9 In Papers 1 and 3, you can decide how long to spend on each part. In Paper 4, you answer as you listen and then copy your answers on to the marksheet afterwards.
10 2
11 It's important to write only the number of words asked for. If you write too little you may not answer the question. If you write too much what you write may be irrelevant. But the numbers are approximate – don't spend all your time counting.

12 Five parts: Multiple-choice cloze, open cloze, word building, gapped sentences, key word transformations.
13 No, only one answer is accepted on the mark sheet.
14 In Papers 3 and 4, yes. In Paper 2, spelling is one of the skills tested, but other things are more important.
15 Twice
16 Five minutes
17 Yes
18 No. You are each marked separately against the criteria.
19 Approximately two months.
20
a) Your grade, your profile telling you which were your weakest and strongest papers and a score out of 100.
b) Your grade

Recording scripts

UNIT 1

page 6, Listening: multiple choice (Part 1), Exercise 2

1

Emma: Next up on the programme, Terry's here to tell us about the London Guitar Show.

Terry: That's right Emma. The London event's simply the largest public guitar show the world's ever seen. And for anyone who's been thinking of taking up the instrument, this is one event that simply can't be missed.

E: Right, so fill us in on what exactly happens at the show, would you Terry?

T: OK, well basically we're talking about a trade show – all the big-name manufacturers will be there – and smaller specialist ones that make wonderful hand-built instruments for the real professionals – so walking round you really can get close to the hottest guitars around, not to mention amplifiers and other gear. Then, on top of that, there's a 3000-seater auditorium where some of the world's greatest players will be popping in to perform – now I mean how cool is that?

E: And any chance of, you know, a bit of hands-on action?

T: Absolutely. These firms want to sell instruments, so they're happy for you to have a go, no matter how well you can play – so there's no need to be shy.

E: Sounds great Terry … so give us a few more details about …

2

Man: Have you noticed how music fills our lives these days? I mean, I'm not talking about the stuff you listen to for pleasure or in the car, but the stuff you hear whether you want to or not when you're out shopping or in a restaurant. There are certain pieces of music I used to like and I cringe every time I hear them now because I've heard rubbish versions of them tinkling away in restaurants, shops or on the phone when I have to wait in a queue.

Woman: I know what you mean – but background music has its place. I mean, who wants to sit in a completely quiet restaurant – especially if there aren't many people there – it covers up the silence. Anyway, I'm quite a fan of background music. I have it on when I'm working on my computer at home. Classical music's meant to be best – Mozart or Bach apparently – it's meant to focus the mind. It's silence which is distracting because you notice all the little noises coming from outside.

M: I don't agree. If you ask me, if you were really interested in the music, you'd listen to it.

W: Not necessarily. Sometimes …

3

Man: So what did you think of it?

Woman: Well, I think the CD's brilliant so I was prepared to be disappointed because these groups can't always get the same effects on stage as they do in the recording studio – you know, you get the noise, the excitement, but you lose the subtlety of the melodies and lyrics. But I couldn't have been more wrong. To my mind, it was every bit as good.

M: Well, I generally prefer to see groups live actually – I really don't go for that kind of sanitised studio sound – it doesn't have the same edge – so I'd have been happier if they'd let go a bit – but you're right, they did stay pretty faithful to what's on the album.

W: And the lead singer's voice is out of this world – I mean how does she hold those notes like that? I thought they could've done some new material though. Apart from one or two covers of rock standards, it was all tried and tested stuff.

M: Oh well, that's what people come to hear – they are promoting their album after all.

W: That's true I suppose – and they got a good reception from the audience generally, didn't they?

M: They did.

UNIT 1

page 13, Exam focus, Paper 5 Speaking: spoken questions (Part 1), Exercise 1

Interviewer: Brita, what kinds of television programmes do you think are worth watching nowadays?

Brita: I don't really watch much television … I don't know.

Interviewer: What about you, Petra?

Petra: I am the same as Brita – I don't watch much television either, but I know that there are a lot of bad programmes! I think reality shows are very bad – I know about them because they are always in the newspapers – you know, celebrity pages and things like that … but my friends like watching sport – good football is worth watching!

Interviewer: Brita, what kind of music do you enjoy listening to?

Brita: I don't have time to listen to music … I suppose I like pop music … but no one in particular. The Arctic Monkeys? The Darkness?

Petra: Oh, I agree with you completely – I like them too – they're great! I also like lots of other kinds of music – and I enjoy musicals like *Phantom of the Opera*. I've got the DVD of that!

UNIT 2

pages 18–19, Listening 1: Exercise 2

It has become a staple of the advice offered by financial self-help books and experts: if you want to save more money, skip the $4.00 coffee you have every day. Compounded at eight percent interest annually, that small sacrifice will be worth more than $70,000 in 20 years. Fine, but what about if you enjoy the coffee? If you were to apply their advice to every aspect of your life, you'd live in a shack, hitchhike to work and only ever eat dry bread. Their advice may help you save, but it won't help you live.

UNIT 2

pages 18–19, Listening 1: Exercise 3

Basically if we look at the ways in which people spend, spenders can be placed in one of three categories. First of all, we have the 'sleepwalkers' who simply do not pay enough attention to how much they are spending. A sleepwalker is always running out of cash, or finds that their bank account is in the red and they can never understand how this has happened. They're not necessarily shopaholics, they just get through money like water, buying things on impulse and never putting anything away for a rainy day.

Then there are the 'status seekers'. They basically just buy things in order to impress others. What drives them is the need not to be outdone by their friends and they want people to see that they've got enough money to spend – in other words, they're slaves to what you might call conspicuous consumption. Any savings they have are there for one purpose: to buy the next 'must-have' item when it appears in the shops.

And finally we have the 'scrimpers': They know how to save alright, but even they occasionally have problems. Like they may go on a shopping spree to offset all their sacrifices or they may be so mean that they can't bring themselves to make the sound investment that would create a nest egg for the future.

UNIT 2

page 19, Listening 1: Exercise 4

Interviewer: So what should people in each of these categories do?

Psychologist: Well, sleepwalkers, for example, need to pay their bills online, to make sure that they pay them on time.

I: So that they don't fall behind?

P: Exactly. And I'd advise the status seekers to write all their cheques by hand rather than using a credit card.

I: Why?

P: It means that you're more aware of exactly how much you're spending at the point of sale.

I: When there's still time to change your mind?

P: That's it.

I: And for the scrimpers?

P: They need to set themselves a budget that includes a reasonable amount of spending. They don't need to economise on everything, they could plan for the occasional treat too.

UNIT 2

page 19, Speaking: giving opinions, Exercise 1

Interlocutor: Here are things which seem to be important in life nowadays. Explain why you think these things seem to have become such an important part of life, and then decide which two you think are not really so necessary for a good life.

UNIT 2

page 19, Speaking: giving opinions, Exercise 2

Man: So what we have to do is explain why these things have become an important part of life nowadays, and then choose two that are not so important.

Woman: Yes, that's right. So let's start with money. Do you think that's more important now than it used to be?

M: Well, yes, actually I do think that. Of course, it's always been important but now I really believe the whole focus of people's lives is making money.

W: I'm not entirely sure that I agree with you there – people have a lot of leisure time nowadays – that's the first picture, and having fun is really important.

M: Yes – but don't you think that we need money to enjoy it?

W: Yes, all right – you've got a point there. But it's not the be all and end all, that's what I'm saying. I just don't accept that it's become so all-consuming.

M: Well, even if I go along with that, it still seems to me to be pretty important. Anyway, let's agree to disagree for the moment, and let's move on. Why do you think clothes are important?

W: It's not clothes, is it – it's fashion. And that's all down to advertising – you know, selling a lifestyle that we all want. And being part of a group, I suppose. Though I don't understand why they think we're all so easily impressed!

M: Absolutely – it's everywhere nowadays – advertising, I mean.

UNIT 2

page 24, Vocabulary 1: compound adjectives, Exercise 4

Speaker 1: I thought that we had a deal – you promised last January to buy the flat, and now you're pulling out. It's unethical!

Speaker 2: We haven't got much time – we'll have to bite the bullet and make up our minds before 6 o'clock – and it's 5.55 now. I think we just have to accept the deal for now, and see what happens. Does everyone agree?

Speaker 3: As you all know, we've been adhering to the marketing plan put in place by the company chairman five years ago, and it has proved very successful – so successful in fact that there is no reason to change our tactics in the near future.

Speaker 4: I heard the finance minister on the radio tonight, and I really couldn't believe what I was hearing – honestly, for someone who is supposed to know about financial strategy he seems to have made a complete mess of things.

Speaker 5: Look – I don't think we can look much further ahead than the next couple of days, but we can certainly follow your idea until we know more.

UNIT 2

page 25, Listening 2: multiple choice (Part 3), Exercise 2

Interviewer: Today we're talking about money and whether having a good salary and a good standard of living actually makes you happy. I'm joined in the studio by the sociologist Graham Styles and by the journalist Sally Greengrass. Let's turn to Graham first, who's going to tell us about some research into this area.

Graham: Yes, a study undertaken in Cambridge established that more than six out of ten people questioned in the city feel they couldn't afford to buy everything they really needed, even though the vast majority, judged by world standards, live lives of luxury. People there earn, in real terms, three times what people did in the 1950s. But the survey shows that they're no happier than they were then. I think this is very worrying and we should be asking ourselves why people have such perceptions.

I: But what does this mean in practical terms, Graham?

G: OK, what it means is that while a £300 fridge will do perfectly well, people actually yearn for a £3000 luxury model. It's what you might call 'luxury fever'. The desire to emulate the lifestyles of the very rich has led to booming sales of luxury cars, professional-quality home equipment, even cosmetic surgery. And the media and our cult of celebrity is partly to blame. It's always been the case that people wanted to keep up with the Joneses, but it used to be that the Joneses lived locally. Now they're the people we see on television – everyone thinks that the celebrity lifestyle is within their reach.

I: But does it really matter? Sally, what do you think?

Sally: I do agree that rampant materialism to impress the neighbours is unattractive, but isn't Graham rather overstating the case here? It seems to me that aspiring to own objects that are beautiful, well-crafted and, yes expensive, is part of the natural human pursuit of pleasure. Owning something aesthetically pleasing that you love and have striven for is satisfying and helps promote well-being. I can't see anything wrong in that, or that people are any different now to what they've always been in that respect.

I: Graham, how would you respond to that?

G: No, Sally's right. I wouldn't deny people the right to have luxuries in their lives. I have a nice laptop and an expensive watch myself. I'm not saying we should go around wearing rags and living in tents or anything like that! The problem with consumerism isn't the objects themselves, but the attachment we have to them. It's that our possessions can end up owning us because we don't really have the means to pay for them. And this does matter because these attitudes are damaging the quality of our lives and damaging the planet too. Credit card debt has trebled in the last seven years and this has been accompanied by a sharp rise in personal bankruptcies. People appear to want everything now and are willing to go into the red to get it, added to which producing all this stuff only adds to pollution and uses up finite resources.

I: Sally?

S: Well, not all luxury goods production pollutes, but it does all create employment and so also creates wealth. The largest luxury goods companies employ tens of thousands of people, and that's without counting the retail sector. When several British fashion houses closed recently, there was a lot of concern because some of the world's most skilled seamstresses and embroiderers were left without employment prospects. Happily, many of them have now joined companies producing deluxe ready-to-wear clothes – companies kept afloat by exactly the prosperous consumers whom Graham despises – people who can afford to buy quality craftsmanship. I'm not sure what sort of world Graham is actually proposing – I seem to remember my parents complaining that the 1950s were rather dull and grey.

I: Graham – a last word from you?

G: Well firstly, I don't despise anyone. But I do think we have to look at the wider costs of rampant consumerism. And I think perhaps the most serious aspect to all this is the damage it does to family life. British parents are sacrificing time with their families in order to work longer hours, and they're doing this so that they can earn the money to keep those families in just the sort of luxuries we've been talking about. And it's all thanks to advertising, to the television and celebrity magazines and all that. So the time parents spend earning money to provide so-called celebrity lifestyles for their children often comes at the cost of those children's emotional well-being.

S: Well, sorry, I think that's really simplistic. Lots of things affect the quality of family life.

I: I'm sorry Sally, that's all we have time for. Obviously this is a topic that'd be worth returning to on the programme, but now it's time to go over to …

UNIT 3

page 33, Exam focus, Paper 4 Listening: multiple matching (Part 4)

Speaker 1: There were three Sarahs in my class at school. Whenever the teacher said 'Sarah', we all turned round. Some people might have laughed it off, but I was embarrassed. I really felt that my name stopped me from standing out in a crowd. Just think – when I used it in my email address, they wanted to put the number 62 after my name. I mean, how cool is that? I scanned the magazines and cast lists of soap operas looking for a new name, but nothing sounded quite like me. Then, one day, someone at work mistook me for a girl called Sian he'd met somewhere. I don't know who she was, but I thought 'Yeah ... that's me ... Sian'.

Speaker 2: Coming here from South Africa, seven years ago, I soon realised how awkward it was having a name, Reimer, that nobody'd ever heard before. I got mildly irritated when people kept forgetting it and then really fed up when they'd keep getting it wrong. When I started working in a large hotel, I knew I'd be meeting lots of new people, and it seemed a good moment to get a new name too. I felt I needed to choose quite carefully though, you know, something that sounded like me. I actually did quite a bit of research – without much success. Then, one day, I happened to hear some people talking at reception and one of them was called Renee. It seemed to fit the bill perfectly.

Speaker 3: My father chose the name Gladys, but I couldn't stand it – it sounded so silly. I mean I wasn't made fun of or anything, but as a child I longed to have a normal name. I started thinking about changing it in my late teens when I was going out with a Chinese boy who found it difficult to get his tongue round. I told him to call me Anne, because we'd just seen a film with a girl called Anne in it and he said it was much easier for him. After we split up, the name stuck. When I started my first job, my new colleagues knew me only as Anne. I loved hearing them say it and that really was the end of Gladys.

Speaker 4: My mother was French and in France, Eric is a posh name. Then, when I was ten we came to England and I soon found the name doesn't carry the same status here. Later, when I was working for a recruitment agency, I noticed how a candidate with excellent experience but a complicated foreign sounding name wasn't getting interviews. I sent firms his details under the name Dave Brown and the interviews flooded in. So when I was made redundant and couldn't get another job, I tried the same thing. I borrowed the name Ethan from a well-known newsreader because looking at him I thought it had the style in England that Eric has in France, and it certainly did the trick.

Speaker 5: When I was born, my parents were going through a divorce so they didn't devote any time to what to call me. My mother made a snap decision that I'd be Sharon, so it lacks any emotional relevance. When I was a kid my brother use to tease me by calling me 'Tiggy' because he thought it was funny. Then when I went to university, I decided to introduce myself to everyone as Tiggy and that was that. But when I started applying for jobs, I worried that nobody would take me seriously if I used Tiggy, so I went back to Sharon. On my first day in a job, though, the first thing I'd say to colleagues was 'oh call me Tiggy please' and nobody seems to mind.

UNIT 3

page 37, Speaking: language of possibility and speculation, Exercise 2

Man: Well, it looks to me as if the young couple are probably quite close – they could be girlfriend and boyfriend. My guess is that it's a long-term relationship because they seem to be very happy together.

Woman: Yes, agreed – that's very possible, but I also think that there's a good chance that they are siblings. I get the impression that they know one another very well, and they look really comfortable with each other.

M: Possibly – you could be right. But I think they probably don't always see eye to eye! What about the young couple then in the second picture?

W: I did think at first that they were colleagues, but on second thoughts they seem to be closer than that.

M: Yes – I think that they are certainly colleagues, but I wouldn't be surprised if they were in a relationship outside the office.

W: The last picture is pretty obvious, isn't it?

M: Well, she's got to be his grandmother.

W: Or aunt ... They certainly seem to be comfortable together and she seems to be helping him with his homework or something like that.

M: I suppose it's just possible that she is a friend of the family – though I think it's unlikely ...

UNIT 4

page 43, Speaking: Parts 3 and 4, Exercise 1.1

Interlocutor: Here are some things that have changed the way we live in positive and negative ways. Talk together about the positive and negative effects these things have had on our lives. Then decide which two have really changed our lives for the better.

UNIT 4

page 43, Speaking: Parts 3 and 4, Exercise 1.2

Man: The first picture shows a person holding some tablets – it looks as if she's a nurse because she's wearing some kind of a blue uniform – that's what nurses wear in the UK, isn't it?

Woman: I think so, though in my country they often wear

white. I can see that she's got a red pen in her pocket, which she probably uses for making notes about medicines. She's holding a cup with what looks like water in it – probably to take the tablets. Medicine has had a very positive effect on life today, hasn't it?

M: Yes, I agree with you, though it has some drawbacks as well. She must be going to give the medicine to someone because she is holding her hand out, offering it to a patient. I can't see her face, though, so I don't know what she looks like. I think she has a watch on her uniform, too – it's important for medicine to be taken at the right time otherwise it can be harmful. What about the next picture?

W: Well, that shows a group of people watching television – I think it's a family because there seems to be a couple and two younger people.

M: I agree with you – and they are watching a very modern television – I think they call it a plasma screen – they're very expensive! It's a modern room, too – there's a big window that looks out on to a garden of some sort.

W: I can see that too – what is it that has had an effect on our lives in the picture do you think? Is it the entertainment?

M: Probably – it's great to have so many programmes on satellite now!

UNIT 4

page 43, Speaking: Parts 3 and 4, Exercise 1.4

Man: Well we are supposed to talk about the positive and negative effects of these different things on our lives – and in the third picture it seems to me that what it's showing is the use of technology – the guy with the mobile phone. He's reading some kind of text and what I guess it shows is the way we can't manage without phones nowadays. That's a real change in the way we live – phones seem to be everywhere – and people use them all the time wherever they are.

Woman: So are you saying that phones have a positive or a negative effect then?

M: Well, I think it's both really – what I mean is, we can communicate with anyone we want at any time, but the phone seems to rule our lives too. There's nothing more annoying than sitting in a train carriage with phones going off all the time!

W: I know what you mean – and I feel that we've lost respect for other people because of them – people just talk loudly and don't think about the impact on others. I was even in the theatre when someone's mobile rang – and they answered it!

M: That's terrible – so impolite! What do you think about the other picture?

W: You mean the girl with the calculator? That's not easy, because although calculators make life really easy I feel that they stop us thinking as well – I saw an article recently that said that older people can do mental arithmetic because they learned it without calculators, but schoolchildren can't even divide 32 by 8 in their heads!

M: So you're saying that they've caused us to become less intelligent?

W: Not exactly – I'm trying to say that they have affected us by making us lazy. What I mean is that like mobile phones they are good and bad!

UNIT 4

page 43, Speaking: Parts 3 and 4, Exercise 2

Student A: Actually, that's a good question. I feel that we do turn to science too quickly – there may be things in life that we can't explain through logic or science. Where's the magic in life? It's all been taken away by people finding explanations for everything. And what's more, science isn't always right! Do you agree?

Student B: I know that scientists get it wrong sometimes – but it seems to me that there is always a reason for things and I want to know what it is! Another thing I think about that is that science has made such huge advances in medicine, for example, and we have all really benefited from those. So no, I don't think that we depend on it too much – I think it's a necessary part of life. And on top of that, I find it really interesting!

UNIT 4

page 45, Exam focus Paper 4 Listening: multiple choice (Part 1)

1

Miranda: Hi Phil, fancy seeing you here.

Phil: Miranda! You're looking great. I wondered if I'd meet anyone I knew at this conference. How are things going at the lab?

M: Well, I'm really enjoying it. I mean, the salary's not great initially, not compared to lots of graduate appointments, but to be honest that doesn't bother me so much. I mean I'm learning such a lot by actually doing things rather than just reading about them and the prospects are there. I'm expecting to move up a grade when my post is reviewed next month. And teaching Phil, any regrets?

P: Well, it's certainly challenging. But fortunately the school's quite well equipped, so I can get the kids doing experiments and take them out on trips and all that, so it's quite fun. But it's really great to come here and get back in touch with what's happening in the world of science – I mean the kids are dead keen, come in with lots of questions and stuff they've found on the Internet, but basically the bottom line is you've got to stick to the curriculum or you're in trouble, so it is good to broaden my horizons a bit.

M: Right. And any news of any of the others …

2

Tom: Basically, we're both scientists and we met in the lab at the university where we both work, so when we thought of starting a family, a job share seemed the obvious idea and we do one term on and one term off which works really well. People are constantly surprised by the size of our family, probably because Caroline looks so young. They say: 'I don't know how you keep so calm.' But it's no big deal really. For us, the biggest hurdle was going from four to five, because George is such a strong character and the wild one! The downside is that you can't make snap decisions; everything has to be mapped out well in advance.

Caroline: We're not unambitious when it comes to holidays though, even if because of the work commitments, we can't often all go together. Nonetheless, I think it's important to share your passions with the kids, so I took the older ones white-water rafting last year, staying in a guest house I found on the Internet, while Tom held the fort with the younger ones at home. They're pretty resilient actually; we had the usual spats and tears, but basically the trip was hassle-free.

3

Maggie: You were late again this morning, I hear.

Ben: Oh don't talk about it. I mean I do try, you know. It's just that some days, it seems that anything that can go wrong, will go wrong. You know, there I was, all ready to leave the house and this chap comes to the door, wanting to read a water meter or something. Anyway, I never even knew we had one – it's in the kitchen apparently, at the back of a cupboard – and as I'm getting the stuff out of the cupboard, you know, as quick as I can, of course I manage to drop a glass jar full of tomato juice – my own fault I suppose, but it broke and went everywhere. I mean, the chap wasn't there two minutes, but I had to stay and clear it up, didn't I?

M: So did you tell Sonia all this?

B: No way. She'd only think I was making excuses.

M: Well, these things do happen, you know. I think it would be better to come clean – otherwise she'll think you're just unreliable.

B: But you're never late, Maggie.

M: I can't deny that – but I have other issues with Sonia and I just think you're better being straight with her, that's all.

UNIT 4

page 47, Listening: sentence completion (Part 2), Exercise 1

Presenter: And now it's time to look at some of this week's new books. The first one I've chosen is called The Red Canary and it's by Professor Tim Birkhead. Many recent books have debated the latest discoveries in the world of genetics, but Tim Birkhead's book is something different. He is Professor of Biology at Sheffield University, and his special interest is the behaviour, evolution and genetics of birds.

If you had to put it in a category, I suppose you'd say that the book is popular science rather than an academic work. It begins with the history of Europe's obsession with songbirds, from the first breeding experiments in medieval times through to the modern science of genetics. Professor Birkhead's style may not be racy, but no one can deny that his research is extremely thorough, as he follows up leads from one ancient manuscript to the next.

But the real focus of the book is one of the Professor's heroes, Hans Duncker, a German schoolteacher whose breeding experiments with birds like canaries and budgerigars in the early twentieth century gave people some of their first insights into the world of genetics. Duncker's great project was to breed a canary, that was not yellow – the bird's natural colour – but red. In order to achieve this, he set out to create hybrids – birds whose parents came from two different species – believing he could transfer the genes that would produce red feathers from one species to another.

He created a new breeding bird with DNA from two separate species, and it is this fact that justifies the rather exaggerated claim, made on the book's cover, which reads: 'The Red Canary: The world's first genetically engineered animal'. But this is hype really because these days, genetic engineering has a very precise meaning that goes a lot further than just breeding animals, and none of that science existed until long after Duncker's death in 1962.

In fact, Duncker never successfully produced a red canary at all because environmental factors, as much as genes, are involved in determining a bird's colouring. Flamingos, for example, are pink, but this is partly because their diet is made up of things like shellfish, which are themselves pink. The canaries, therefore, needed to be eating food that contained what are called carotenoids if they were to go red, and Duncker didn't discover this.

Having enjoyed it all the way through, the book's ending came as a bit of a disappointment to me. Like many other scientists, Professor Birkhead tends to judge the past by the standards of today rather than those of the time. Nonetheless, it's a good read – highly recommended.

UNIT 5

page 54, Listening 1: multiple matching (Part 4), Exercise 2

Speaker 1: There are so many more gyms and things like that now – we used to go running round the streets but now everyone goes to the gym so that they can use the latest equipment!

Speaker 2: It seems to me that everyone just wants to make money out of it these days. And everyone wants to win. When I was at school we played for the fun of it.

Speaker 3: It's become such a big thing these days – everyone is completely obsessed with diet and exercise – and, of course it's become big business too. You seem to hear about nothing else on the television. I think it's too much really.

Speaker 4: I love the way these sports that you've never even heard of keep popping up – almost every week – you know, boxercise, things like that – some people think they're not really sport at all, but I say why not? In my day all we had was tennis!

Speaker 5: Well – how much did that footballer get when he signed that advertising deal? It's ridiculous what some of these people get just for being who they are! I can't see what it's got to do with the sport really.

UNIT 5

page 54, Listening 1: multiple matching (Part 4), Exercise 3

F: What do I think about sport changing in the future? Well, I think it really depends on whether you're talking about people who actually do it or people who watch it – and whether you're talking about elite sportspeople or those who do it for fun. To be perfectly honest, I'm not sure whether the world of professional sport will really change much in the next few years.

M: Really?

F: Well, professional sport is very conservative at heart – I mean, think about all the so-called 'great' sporting events, like the World Cup, or Wimbledon, they're only great because of their long history – it's the event not the quality of the sport itself, you know, being part of that great long tradition. World Cup matches can actually be very boring!

M: True! I've seen some deadly ones in my time! So you think that, broadly speaking, we'll be watching the same mainstream sports in the future as we do today?

F: Yes, I do. But that's not necessarily the same for those who play sport – the amateurs. That's us! We tend to live in cities, and that's got to have an influence on the kind of sport people can do – or have access to. Any game that works well in confined spaces – such as basketball or five-a-side football – has an advantage. So other games that people play on Saturdays like, er, cricket or baseball, may well start declining; that's because fewer people will be able to play these games because of the space that the playing area takes up.

M: Good point. And you're right about new sports coming along – my friend does 'Bungee Running' – where you stretch a piece of elastic until you're 'pinged' into a safety net ...

F: ... and I know someone who does 'Bouncy Boxing' – where you hit a friend with giant inflatable gloves. Then there are other things that are less 'off the wall' but have started as a result of the ever-increasing interest in health and fitness – I think that's a trend that will carry on. I mean things like Boxercise, which is a combination of boxing and physical exercise. Though, as I've already said, that doesn't mean that I see the total end of more traditional sports.

M: No, you're right – we'll always watch traditional sports, and I think that whatever happens, people will still have their sporting heroes. Television has made sport big business in the last 30 years – sponsorship and so on – that'll keep going, I'm sure, and soon the sporting elite will be just like film stars – they'll be paid even more and will do even fewer performances! I wish I could be one of them!

UNIT 5

page 59, Speaking: agreeing and adding information (Parts 3 and 4), Exercise 2

M: So what do you think about this one? I mean the idea of having yoga – or is it just keep-fit? – you know, classes for young people?

F: I think it's great – it's good to let young people try lots of different things, things they wouldn't otherwise do – and on top of that it's obviously good for their health! What's more, they won't get bored because they're doing it in a class and they have friends around them so they're more likely to keep going. The teacher looks really interested in what she's doing, which is good.

M: I take that point on board – but having said that, I think that it's hard to do a lot of the moves in some keep-fit classes – and certainly in yoga! – this picture looks difficult to me – and they might get put off if they can't do everything. Not only that, but it's rather non-competitive – they might just think what's the point really? So what about just building more leisure centres? Like this one, with a gym?

F: That's all very well, but we do need to remember that that would cost a lot of money – is it actually cost-effective?

M: You could be right – So let's move on ...

UNIT 5

page 63, Listening 2: multiple choice (Part 3), Exercise 3

Interviewer: Today we're looking at the sport of indoor climbing and with me here in the studio I have the American climber Tom Lake and the London-based sports journalist Amy Styles. Tom, indoor wall climbing in the USA is on the up, isn't it? What type of person does it appeal to?

Tom: Well, according to a recent survey, nearly nine million Americans now go indoor climbing each year, and even if the number of climbing gyms in the US doubled, we reckon it still wouldn't have reached saturation point. Indoor climbers are different from traditional climbers. The style is more explosive, more athletic. And the sport seems to be particularly attractive to 13- to 21-year-olds. Maybe that's because it feels more egalitarian than outdoor climbing, with the regulars in most gyms offering advice and encouragement, rather than trying to outdo anybody. I mean, there's no race to get to the top in indoor climbing.

I: But there are games you can play on the wall, aren't there, Amy?

Amy: Oh yes, games on the climbing wall are fun and create interest and usually extend the length of the workout. For example one popular game is called 'add on' and basically two climbers of similar ability – you know, it could be two novices or two old-hands – begin by agreeing on the first sequence of moves. The first climber gets on the wall and climbs this sequence of moves. When he finishes, though, he then adds one more move to the sequence, so the next person has a bit more to do, and so on. You mark the moves with chalk or memorise them if that's part of the game. It's a great game for learning sequences, and that's the real point of it actually because that's something all indoor climbers have to do.

I: So what type of people are we talking about in London, Amy?

A: At the climbing wall I go to in London – which I think is fairly typical – the climbers are mostly of student age. And let's face it, climbing indoors is cheaper and less time-consuming than venturing outdoors at weekends with expensive equipment, and that's a big part of the appeal. To me they seem to be, like, the well-educated big brothers and sisters of the inner-city teenage skateboard crowd – it kind of picks up on that scene and all that goes with it. So it's quite a rough-and-ready environment really, and the atmosphere is also kind of sociable – but people take the climbing pretty seriously for all that.

I: But can climbing up a wall really be compared with climbing up a rock face, Tom?

T: The lack of natural rock isn't necessarily a problem. I mean it's not a blank wall, there are plastic hand and foot holds and plastic obstacles to negotiate too. You're never more than a couple of metres off the ground, but even at that height the focus on staying glued to the wall can be intense. As any climber will tell you, just because it's indoors, doesn't mean it's tame. And if you're really committed, it can be just as arduous. There's the fear, however irrational, that you might get hurt, even though there are crash mats below you.

A: Yes because you've also got to remember that it's also a good mental workout, forcing you to solve all sorts of problems while striving not to fall off. You need a lot of core strength – so it's good for stomach muscles and for arm and leg strength – but 'I was OK until I started thinking' is a common complaint amongst indoor wall climbers, because the challenge is as much mental as physical. And that's another reason why it's become so popular, it demands far more than the usual gym-based workout. On top of that, regular climbers develop an enviable lean physique; second only to surfers really.

I: Tom, you climb both on walls indoors and on mountains outdoors. Which do you actually prefer?

T: Well, to my mind indoor climbing is every bit as valid as the outdoor variety and I really can't go along with people who say that the only real climbing is scaling up the side of Mount Everest or wherever. But having said that, for me you can't beat outdoor climbing, perhaps because that's what I came to first, who knows? But I don't feel any need to put down the indoor sport in any way – why should I? It's just a different experience and I get a great deal of enjoyment from both.

I: Tom, Amy, thank you very much for joining me today.

UNIT 6

page 69 Listening 1, Exercise 1

1

M: We're rather strict parents, so we believe in things like discipline and good manners. People are always rather flabbergasted at this – wondering how we manage it against the background of all the stuff they see on television and the influence of their peer group. But actually it's harder for the kids.

F: Yeah, because they know what we expect of them, and when they're with us, that's fine. But when they see other kids running riot and getting away with murder, it's almost as if they're embarrassed.

M: There's nothing wrong with being boisterous and loud when the moment's right, but the real social skill is picking up on what's going on around you and the effect your behaviour might be having on people.

F: I like to think ours have that.

2

Presenter: Karen, you mentioned babysitting …

Karen: That's right. The rows started about the babysitting. I'd been used to having little brothers and sisters bombing about, and being the eldest I often ended up keeping an eye on them when Mum popped to the shops or whatever. But then when they went out anywhere, my parents would pay this girl to come and sit with us – I mean she was only a couple of years older than me. So, I put it to them, you know, 'pay me – I'll do it for half the price'. Then I got this long spiel about babysitters having to be 16 and mature for their years and all that. Well, I'd checked that out on the Internet and I couldn't find a trace of any such law. Anyway, I'd got the message alright. From then on it was war.

UNIT 6

page 71 Exam focus Paper 5 Speaking: collaborative task/discussion (Parts 3 and 4), Exercise 1.1

Examiner: Now I'd like you to talk about something together for about three minutes. Here are some pictures showing different things that can have an impact on family life. First, talk to each other about the positive and negative impact on family life reflected in these pictures. Then decide which picture best shows the biggest impact on family life today. All right?

UNIT 6

page 71 Exam focus Paper 5 Speaking: collaborative task/discussion (Parts 3 and 4), Exercise 1.2

Pascale: So our task is to discuss the positive and negative effects of each of these things on family life – is that right?

Fernando: Yes, and then at the very end we have to decide which one has had the biggest impact.

P: But we mustn't do that first – before that we have to discuss each one. So, what do you think about this picture? I mean the one with the new baby.

F: Well, it seems to me that a baby has a big impact on family life because it changes the amount of time parents can give their other children. A baby upsets the balance.

P: So what you mean by that is babies are very disruptive to family life – is that what you're saying?

F: Yes – though I haven't got any children myself yet.

P: No, nor me! But I'm sure you're right. Let's move on to the girl with her L plates – she's passed her driving test, so she has got a lot more independence now – that can cause a real change in the dynamics of family life. Parents lose some control over what their children do! What I mean is – it can be the start of the children leaving home, and that can be difficult. Do you feel the same as me about it?

F: Yes, I suppose so, but she looks happy about passing! It's good to be able to drive when you're young, I think.

P: But do you agree that it can have a big impact on family life, or don't you think it has as much effect as a baby does?

F: I'm not sure – I suppose it's not so big really, though it can alter the way families interact – after all, the parents won't have to drive her everywhere now. And she will feel more confident because she has passed. She might not spend so much time with her family.

P: That's true – it might give her parents an easier life, though the relationship will be different. Let's discuss the next picture …

UNIT 6

page 76, Listening 2, Exercise 2

Gaynor: We have very old-fashioned, traditional values. My husband Rhodri is the breadwinner. He works more than 90 hours a week on the farm and does nothing at all in the house. I can count on the fingers of one hand the number of times he's changed our son's nappy and he's never changed the girls'. He might make a sandwich at a push, but he's never been known to make a meal. Ever. The children are always deliriously happy to see him when he comes in from work, but he doesn't spend any time with them. He's too tired.
I'm the anchor at home. Rhodri would be appalled if I went out to work full-time, and I'm happy to be a mum and farmer's wife. I think there is no substitute for mum bringing up the children. Megan is a real tomboy. Delana, my two-year-old daughter, is very girly. She likes pink, sparkly things and dolls. Megan likes dirt and puddles and being pushed around in the wheelbarrow. We live in the middle of a 14-acre field, so there's lots of space for her to run around in. If she's given a choice of dolls or diggers, she'll choose the mechanical toys every time. Research has shown that boys' longest fingers are their ring fingers. For girls, the index finger is usually longer. However, tomboys have longer ring fingers. Significantly, Megan's ring finger is longer than her index finger and she certainly shows all the signs of preferring boyish toys and activities. She loves building with Lego™ and she already helps on the farm. We'll say, 'Megan, hold this', 'pass that', 'fetch that', and she'll rush off happily. We don't have holidays or buy many clothes. The children have second-hand stuff, usually leggings and trousers, and Megan tends to wear boiler suits and jeans. Neither of the girls has pink clothes. I'm not a pinky person. I won't buy them Barbie dolls, either. I'd rather they played in the sandpit or with modelling clay or did some drawing.

UNIT 6

page 76, Listening 2, Exercise 3

Marie: I didn't want a boy. When I was pregnant with Tyrese I said, 'If it's a boy I'm leaving him in the hospital!' Boys become men, they get into trouble. I already have two girls and I know how to deal with them. Boys are a totally different ball game. I've got five brothers and they all got into trouble in their teens, mixing with the wrong crowd, getting into fights. We were brought up with different foster families after my mum went back to Jamaica when I was 12. My brothers reacted adversely to our chaotic upbringing, but I was determined to do something positive with my life. I believe in self-advancement, the work ethic and I want to instill it into my own kids. I want them all to be successful, to earn plenty of money, to want for nothing. My husband and I believe in equality – we share the household chores – and I'm determined Tyrese won't be a stereotypical male. Women who mollycoddle their sons turn them into awful husbands. As soon as he's old enough, Tyrese will be ironing his own shirts. I haven't treated Tyrese any differently from the girls, but he's different from them. I used to throw the girls up in the air, but they screamed, so I stopped. I do it to Tyrese and he loves it. He's more of a risk-taker, more aggressive, more adventurous than the girls, and he started walking earlier. I haven't bought any boys' toys especially for him, though. He has what I bought for the girls, and I didn't buy them girlie things, anyway. I haven't treated him any differently from them. My husband wants him to be more masculine because he's afraid he'll be picked on at school if he's too soft. I think it would be nice if he had a sensitive, emotional side as well as a masculine side.

UNIT 7

page 80, Exam focus Paper 4 Listening: sentence completion (Part 2), Exercise 1

Presenter: Welcome to this evening's edition of Insight – your weekly look at what's on locally. Well, Saturday sees the beginning of this year's arts festival – the city's twentieth in as many years – and you'll have probably seen preparations for the event going on around the city. You may even have spotted the odd multi-coloured cow popping up in unexpected places.
No, I'm not going mad, the subject of this art feature really is cows – a whole parade of them to be precise. This year, our city is helping to host the world's largest public art event, which is called The Cow Parade. From Saturday, around 100

life-size cows, made out of fibreglass and decorated in all manner of colours and patterns, will be on show at indoor and outdoor locations across the city,

The Cow Parade was the original concept of a man called Walter Knapp who worked, not as an artist in fact, but as a window dresser in Zurich in Switzerland. The cows were designed as a way of attracting customers to shops, but also as a way of promoting the work of local artists whose work was, in effect, exhibited on them. The idea was such a success that the cows have since been used in a number of cities worldwide, including New York, London and Sydney, both to generate interest in the arts, and also as a way of raising money for charity.

Each cow is sponsored by a different company – in our case, for example, by the local art gallery, a travel agency and the city's zoo – to name but a few. What this means is that the company has effectively paid for the materials and the cost of installing the cows which will appear in places as diverse as shopping centres, the hospital foyer and the football stadium. Then, at the end of the event, it is hoped to sell as many as 80 percent of the cows at an auction sale, with 75 percent of the proceeds being given to charity.

In past cow parades, celebrities such as J.K. Rowling, Elton John and Nelson Mandela have bought cows at auction, and the best price ever reached was 125,000 euros for a cow called Waga Moo Moo, sold in Dublin in 2003. Now you might imagine that this cow was gold plated or covered in diamonds, but actually it was decorated with 15,000 tiny bits of glass. Now, there's a record for our local artists to try and break!

This city's cow parade will benefit two large charities – the first is one called VET AID which helps farmers in poor countries to keep their animals in the best of health, which seems very appropriate really. The other is a local children's charity. In fact, hundreds of kids are going to be attending workshops where they'll be helping to create a mosaic design to display on one of the cows using thumbnails downloaded from the Internet. Meanwhile, outside the Central Library, any passers-by will be encouraged to help colour in what's called a 'weird and wonderful flower design' stencilled on one of the cows, in return for a small donation.

If you'd like to contribute to either charity, but haven't got space for a whole life-sized cow at home, ...

UNIT 7

page 86, Grammar 1, Exercise 4.2

William: Sarah, why are you going to Simpson's exhibition – you don't like art!

Sarah: True – but I reckon there's a chance that this could be the beginning of something big – most people won't have heard of this young boy yet.

W: Right – and who knows what he's going to achieve in the future.

S: Exactly! Not that I really know that much about art, but I'm on the point of actually buying one of his paintings.

W: That's a bit extreme! Why would you want to do that?

S: I think it's an investment – in 20 years' time he will have become the most popular artist in this country.

W: I see what you mean – so imagine what his paintings will be worth then!

S: You see my point! So then I will sell my investment for a fortune – and then I won't be wasting my time working in an office – I'll be sunning myself on a beach somewhere on the profits!

W: Good plan – I'll join you! £50 invested now will be worth so much more in 20 years' time!

UNIT 8

page 92, Listening 1, Exercise 1.4

You often see the term 'self-starter' in job adverts – but what do companies mean when they put this? Are they looking for someone who will come in and aggressively reorganise the office, upsetting everyone and interfering in things they don't really understand? No, of course they aren't.

What companies are looking for is someone who's able to work without constant supervision; someone who'll quickly understand what the job demands and quietly get on with it, without someone else needing to check everything they do. That means, of course, someone who can work independently; someone who doesn't need to keep asking questions, but also someone with the common sense and good judgement to ask for advice and help when it's really necessary.

Also, the term 'self-starter' implies someone who's not just going to do the job, but someone who's also going to develop it in some way: for example, find more customers, or find ways of doing things more effectively or efficiently.

UNIT 8

page 92, Listening 1, Exercise 2.2

Well this questionnaire tells you how independent you are, or might be, in a work situation. Basically, people scoring 12 or more are self-starters. They like to be in control of what they're doing – they look for advice rather than supervision. They're people who don't always follow the rules, which can be a bit of a drawback for companies, but they're also the people who find new ways of doing things. Most companies are happy to employ a few people like this – not too many, or the result could be chaos!

People scoring between six and 11 are moderately independent. They like to manage their own time and work with minimal supervision, but they're more likely to fit in with accepted methods and procedures. Companies like to employ lots of people like this; they're relatively conformist, but tend to be open to new ideas too.

People scoring five or below are generally happy to accept supervision and like to work within clear rules and guidelines. They don't question the way things are done, but just get on with what they have to do. All companies need some people like this – they keep things running smoothly and are usually very reliable. The problem comes when too many people in an office are of this type – because things then never change!

UNIT 8

page 97, Grammar 1: direct and reported speech, Exercise 5

1 I was wrong to get angry.
2 I earned more than ever last year.
3 I've never met him before.
4 Please think about what you are doing!
5 I will work harder next week.
6 I believe that overall performance will improve if we give bonuses to our staff.

UNIT 8

page 98, Grammar 1: direct and reported speech, Exercise 6

I had always wanted to work for myself. I never really thought it would happen. Then someone asked me to write a story for the local magazine. Because I found it really easy, I decided to write another one. One thing has led to another, and now I am writing full time, and I love it!

UNIT 8

page 98, Listening 2: multiple choice (Part 3), Exercise 3

Interviewer: My guest today is Amy Kyme who spent four years on the cast of a well-known television soap opera, but has now moved on. Amy, welcome.

Amy: Hi.

I: Tell me. You were 18 when you got that part. Was there a lot of competition for it?

A: Oh it was a complete fluke. I went along to what's called an 'open audition' to give moral support to one of my mates who was going after a part. I mean 40,000 girls across the country went to these auditions, and it was a long day; lots of hanging about. So I got bored waiting for her, and thought as I was there I might as well give it a go myself. I was stunned when, a few weeks later, I got a call offering me a part; I mean it's not as if I'd had any training or experience. But I felt proud to think they wanted me and never considered turning it down. I mean I'd have done it for nothing at that point, though actually they were offering £300 per episode, so if you were filming all week, you could make £1,500. It was good money.

I: So, while your mates were all heading off to university, you moved away from home to start a whole new life as a TV star.

A: That's right. And initially, it was really glamorous and exciting. I ate in the best restaurants, hung out with celebrities, and was on the guest list at all the top clubs. I even bought myself a smart flat in a trendy district. And the work was hardly difficult. Sure, we worked long hours, but I had no trouble memorising my words – acting isn't exactly brain surgery! But it wasn't long before the glamour started to wear off. The cast were all young people with plenty of money, determined to live it up, but I soon lost interest in clubbing and going out. I felt a bit alienated from them really. It's not that I argued with anyone or had enemies, more just that we didn't have that much in common.

I: So was that why you quit?

A: It was more than that. I suppose I was missing my family too because I started comfort eating, and my weight shot up. I was shocked when the show's producer told me to lose weight, though. And I started to wonder whether I really wanted to be in a job where I was judged on my looks. That's what decided me. Then the plotline in which my character, Laura, became dangerously obsessed with something dragged on for nearly two years, and the scripts got pretty repetitive. I'd be reading my lines, thinking: 'Haven't we already shot this scene?' I was getting bored with it. But by then I'd made up my mind to leave.

I: Was that hard to do?

A: Well, I told the producers I wanted to quit to go to university. It was difficult, but as my contract was up for negotiation anyway they agreed to let me go. My family and boyfriend Joe were completely behind my decision to leave the show, and I never doubted that I was making the right choice. After I handed in my notice, though, I couldn't stop worrying about how I was going to pay the mortgage on my flat. I knew I was about to lose a really good salary. The only thing that kept me going was knowing I'd be going on to do something more interesting.

I: So had you already decided what?

A: Not yet. In those last weeks of filming, I'd spend my evenings researching universities and courses online, before collapsing into bed, completely exhausted. After I'd shot my last scene, it was frightening being out of work – not knowing what the future held. So in the end, I decided to sell my flat and moved in with friends. Then, out of desperation for something to do, I took a job as a care assistant in an old people's home. It was a revelation. These pensioners completely depended on me for everything. I loved the sense of responsibility, and the fact that I was finally doing something worthwhile with my life.

I: And that is?

A: I'm now studying midwifery at the university hospital where I work – which I'm thrilled about. My life couldn't be further away from my old soap existence. Most of my days are spent studying or working, and I hang out with my mates in my spare time. I also got married to Joe last July. Finally I feel like myself and not just a character on the TV. Occasionally I see some of the actors I used to work with for a quick catch up, and unfortunately I do still get recognised occasionally, but I hope that by the time I'm fully qualified as a midwife, everyone will have forgotten what I used to do.

UNIT 9

page 109, Listening: sentence completion (Part 2), Exercise 2

Presenter: Next on the programme, we have some news about Tom Kevill Davies, a cycling enthusiast who's undertaking a gastronomic tour of the Americas by bike. The trip will cover 20,000 kilometres and 3000 meals, and what Tom eventually hopes to do is to raise £100,000 for charity in the process.
Tom is 27 years old and when not cycling, works as a graphic designer in London.
But currently he's in central America, having cycled across North America from the eastern seaboard, taking in parts of Canada, then cycling down the west coast of the USA and on through Mexico.
The idea for the trip came when Tom was on a cycling holiday in France. He had a great time, meeting lots of people along the way and enjoying some excellent meals. That's when he thought of doing a long cycle ride for charity, taking in the local food specialities along his route. Tom has recruited a number of sponsors through his website, The Hungry Cyclist dot com, where you can also suggest the types of food that you think Tom should sample on each leg of his journey.
Although it has 14 gears, which are encased in a special box to keep them free of dust, Tom's bike is a fairly standard model. Tom has a portable computer with him to keep in touch with his website, which has a folding keyboard, but apart from a digital camera, that's all the technical equipment he has with him. The bike's one really distinctive feature is a bell in the shape of a burger. A nod towards the local cuisine that he thought would be appropriate for the North American part of his trip.
Tom says that hills and headwinds are a problem, but that punctures are his biggest headache. He's also sometimes slightly frustrated when, because all his stuff's onboard, he can't just leave his bike unattended while he goes off exploring in the places he visits. That's why he's so grateful to all the people who've given him meals and hospitality as well as a chance to rest up along his route.
Sometimes Tom's stops are planned. People send him an email, having seen him featured on local TV or newspapers, and invite him round for lunch or dinner, but others are not. One of Tom's best meals to date came after he'd dropped his gloves as he cycled through a small lakeside community in Minnesota. The man who came running after him with them turned out to be a local restaurant owner.
Although he tried various local specialities as he cycled across North America, Tom remains most impressed by the fruit pies that he came across in small roadside diners almost everywhere. Tom has tried eating wild moose in Canada and ants' eggs in Mexico, but says his most unusual dish to date has been snapping turtle soup.
The animals themselves don't look very appetising, but according to Tom the flavour is not unlike that of chicken.
Tom's hoping to try even more exotic dishes as he heads down through South America. So if you've got any suggestions …

UNIT 9

page 110, Grammar 2: emphasis (cleft sentences with *what*), Exercise 4.2

What I really need to do to improve my English speaking is stop worrying about making mistakes. I know that I just think too much! What I'll have to do in future is respond more spontaneously, and think about the person I'm talking to and not just focus on myself – after all, communication is the most important thing, isn't it! There's no point in trying to get everything perfect if I keep hesitating and feeling nervous. What I'll have to practise is relaxing more and just being more natural!

UNIT 9

page 111, Speaking 2: individual long turn (Part 2), Exercise 3

In picture B the people seem to do this travelling every day, and they appear to hate it. They're just doing things like reading and sleeping because they are trapped in the commuting treadmill, and have no choice. In picture C they are also in a work situation in the same way but this time they are on an executive jet and they are obviously working together while they're flying. What I think is that the people in picture B are just fed up with the whole thing, because commuting is such a waste of time whereas the people in picture C are using their time profitably and for that reason they feel a bit more positive about travelling.

UNIT 9

page 114, Writing: competition entry (Part 2), Exercise 4

It looks great fun! You have to start with your photo, and don't forget to give a bit more detail on what you can see in it, and write down your ideas clearly. I don't think you have to give facts and figures, though. It says that you've got to increase awareness of the issues for readers – that's OK, because you'll be doing that when you explain why you chose your photo – oh, and engage them – that means you've got to use lots of interesting vocabulary. Mind you, you've only got up to 260 words, so you can't give too much unnecessary detail. I think the most important thing is to give your own opinion, really, and last but not least you have to try to win – after all, it is a competition!

UNIT 10

page 116, Listening 1, Exercise 1

1
It's something that everyone seems to be talking about nowadays – every time I turn on the television or open a paper there it is! We have to use less energy, recycle more, protect tigers – I can't be bothered with it really. It's such a hassle! And I don't really know what all the fuss is about – whatever anyone does now is a waste of time because any damage we've all done is probably irreversible. So what's the point?

2
It's arguably the biggest problem my generation will have to deal with – especially as the future of the planet depends on it. And although the information is everywhere – I mean they do a good job in making people realise what's going on – there are too many people with a 'can't be bothered' attitude. They ignore the facts and bury their heads in the sand. It's up to all of us to do our bit, on an individual basis, and I think that everything I do, however small in itself, contributes to something that is so important for us all.

UNIT 10

page 120, Exam focus, Paper 4 Listening: multiple choice (Part 3), Exercise 1

Interviewer: Today's guest is Jake Willers, who's general manager of a wildlife park in England. On a day-to-day basis he looks after animals like emus and tigers, but Jake's life-long passion is for somewhat smaller creatures, and that's why you can now see him presenting the TV programme called *Insects from Hell*. How did this passion begin Jake?

Jake: Because I'd been round insects, invertebrates and arthropods from a young age they never worried me at all. My mum had a tarantula spider when I was five and I had a pet scorpion when I was 11. And they're fascinating. I mean, take arthropods – by which I mean things with jointed limbs and bodies and a tough outer skeleton – there's about one million named species on the planet at the moment but scientists predict that there's possibly over ten million more to find, so we've only just scratched the surface. That's exciting – it makes me think what an incredible area to look in to.

I: But getting the public interested in creepy-crawlies can't be that easy?

J: Basically, in terms of the public perception, Hollywood's done for insects what Little Red Riding Hood did for wolves. You get all these blockbuster films about killer bees and spiders on the rampage. But it makes me laugh when they use tarantulas as dangerous spiders in the movies because they're one of the least dangerous spiders there are. Yes, they're big and hairy and have venom, but it's no worse than a bee or wasp sting. But having said that, the programme I present is called *Insects From Hell* because absolutely anything with '*from hell*' in the title gets good viewing ratings. It gets them watching, I'm afraid.

I: So how did your career as a TV presenter start?

J: Well, a production company came to do some filming at the wildlife park and got me talking about the place. I don't think what I actually said impressed them so much as the fact that I seemed to have an easy, unselfconscious manner on camera. Out of the blue, they pitched an idea for a series. A major wildlife channel liked the ten-minute promo video we shot and commissioned a six-part series to be filmed around the world. It's now broadcast in 150 countries and I've even got a fan club, mostly made up of schoolchildren. I think they like the yuk factor – because I go to some pretty unusual locations like caves full of spiders and dung heaps, and I'm prepared to get my hands thoroughly dirty.

I: But there's more to the series than sensationalism, isn't there?

J: Oh, right from the beginning I said I wouldn't do it if it was just for entertainment. We work with specialists because, although I've got a certain amount of experience, I can't know everything about all the individual species. And it's important to get the facts right because these animals are a vital part of our planet's eco-system and, as such, deserve our attention and our respect. So basically the way it works is the producer asks me to do something. I'm open to anything! I'm into action sports – I sky dive and scuba dive and I've been around wildlife all my life, so I usually say 'yes'! Then they get a script researched and written around that.

I: Didn't one of the highlights of *Insects from Hell* involve something a bit larger though?

J: That's right, and it also brought home to me how much we rely on the judgement of people who really know what they're doing. We were in the Kruger National Park in Africa looking for elephants. We wanted to film the flying beetles that live around them. I was in the four-wheel drive with a chap called Leo, who's the elephant expert, followed by a van with the equipment and crew. Eventually we see elephants crossing the road in front of us and decide to head back down to where they came from. But this young bull elephant at the back has other ideas. He turns round, stares at us and then charges. He's a baby, but he's still twice the size of the car. The incredible thing is, Leo just sits and waits, then puts the car in first gear and drives straight towards it. Suddenly the elephant slams on the brakes, turns round and hotfoots it off after the herd.
Now that was cool, because if it had been an adult, it would have come for us.

I: So Jake, where does your career go from here?

J: It's easy to feel very flattered if someone says: 'We'll make you a TV presenter,' but the park's still my priority. I have a great life. I did 20 different jobs before I came here. I've put aircraft parts together, worked on a fruit-and-veg stall, trained as a chef – you name it. It makes you realise how lucky you are when you get to do something you really love. Doing this and getting to travel the world, and get paid for it as a job, what more could you wish for?

I: Jake. There we must leave it. Thank you …

UNIT 10

page 123, Speaking: sounding interested, Exercise 1

Man: I really think that zoos are an important issue today – it seems old hat but really they are doing a vital job in preserving species.

Woman: Really.

M: Yes, don't you think so?

W: Well, if you do.

M: But just think about the research that goes on – who would do it if zoos didn't? No one else has the time or the interest, do they?

W: Who knows?

M: Don't you think that it's important to sustain wildlife and look after endangered species?

W: I suppose so – I hadn't really thought about it.

UNIT 11

page 138, Listening: sentence completion (Part 2), Exercise 3

Presenter: Hello, I'm Tom Membury and in today's programme we're looking at the role of laughter in our lives, and particularly at one form of laughter therapy that's becoming increasingly popular on both sides of the Atlantic.

The benefits of laughter are well known. If you're feeling down or stressed out by work or study, there's nothing better than a good laugh to help you wind down and forget the stresses and strains of daily life. Laughter then is like crying, you give into it and afterwards you feel a sense of relief as pent-up emotions are released. This is why many forms of entertainment involve laughter, it's why we enjoy comedy films and TV programmes. And as all theatre actors and comedians know, laughter is catching. A few funny lines are all that's needed to get a good audience laughing and once they start, they'll laugh at anything.

The idea of laughter therapy, or laughter yoga as it's known, originated not in the USA as you might imagine, but in India. Dr Madan Kataria, a doctor, read a magazine article that suggested that laughter was better than medicine and decided to put this idea to the test. He began taking a group of five people to a park and got them telling jokes. It soon became apparent, however, that most jokes are offensive in some way and as the group grew in size, he developed strategies to make them laugh without jokes. What he discovered was that laughter helps people to relax even when there is nothing to laugh about; that the physical aspect of laughter is good exercise that relaxes the body even without the stimulus of humour.

The idea of laughter yoga was born.

Dr Kataria has since made a number of claims including, as you might expect, that laughter reduces stress and tension, but also that it boosts the immune system, thereby reducing the chances of minor ailments such as colds and flu, as well as being a good workout in its own right.

Dr Kataria's book entitled 'Laugh for No Reason' has become the guiding principle behind laughter yoga classes all over the world, and an award-winning film called 'The Laughing Club of India' has been made about him.

He makes regular appearances at events such as World Laughter Day in California.

I went along to a laughter yoga session to see what it would be like. I'd been warned that I'd have to take part in clapping and chanting games, but arrived to find the participants walking round the room talking to themselves loudly. In this game, I learned they were talking into imaginary mobile phones. I joined in and, sure enough, every so often one of us would break out into uncontrolled laughter. We then played other games including one where we had to pull funny faces at each other. I did find myself laughing along with other members of the group. And yes, I did enjoy myself.

So what can we learn from …

UNIT 12

page 148, Listening: multiple matching (Part 4), Exercise 3

Speaker 1: This game's awesome. I had hours of fun, and I reckon it's because the developers of the series really know what they're doing. I mean, this one easily lives up to the reputation of its predecessors. Single players go off on a quest as a 17-year-old cadet intent on saving his sister from an invasion of mysterious life forces. Meanwhile online players create their own characters, then set off to explore all three planets in a distant solar system. The usual sort of stuff, but all really well thought out. There's only one downside which is that there's a monthly subscription if you want to play online. Fortunately the off-line part never gets boring, so you can just stick with that.

Speaker 2: It's one of those games that had the unfortunate luck of being pretty much perfect the first time around. What this means is that, apart from technical updates to graphics and sound quality, the developers didn't really have to change much for the sequels. Not until this one that is, and now they've almost ruined everything by over-complicating it. Having said that, a lot of the slightly tedious elements have been cut out, so the whole thing is a tighter, quicker-paced game which you have to give them credit for. What you get is the familiar rich, ever-evolving world that'll provide hours of entertainment, but this certainly wouldn't be the one to start out with as a beginner.

Speaker 3: As a long-time player of the games in this series, I was expecting great things of this new one which has a new cast of characters, but a familiar setting. There'd been a lot of hype with the creators winning awards and all that. But, all in all, it's a bit of a letdown. The graphics, never the strong point of the series, are now kind of OK, but to be honest I was expecting better. If it wasn't for the really deep and long storyline – it's actually interesting and not just something to click through like in lots of games – there'd hardly be anything to recommend what is really quite an expensive game.

Speaker 4: Like a lot of people, I'm a little disappointed having waited for this release which is meant to represent a decade of development experience. Well, I'm finding my game freezes up every time I start to trade with anyone, and sometimes for no reason at all which is a bit puzzling. The main plus point is the fact that the action takes place in Africa, which makes a nice change if, like me, you're a bit fed up with the usual goblins and other typical characters found in games. The plot sounds complex, but actually once you get into the game, it's OK. On balance though, I'd say don't spend any money on it till they've ironed out the glitches.

Speaker 5: Although there's numerous worlds to traverse and dungeons to delve into, the great thing about this game is there's nothing complex about it – you can pretty much progress at your own pace without needing any particular expertise. But don't forget, it's still basically a cartoon, the world you're in is stylised, so you're not seeing complicated settings or the lines on the characters' faces. There's plenty of detail though, even in regular TV resolution. And whilst the audio quality's fine, this game's one of a series which is well known for the silly noises the characters make instead of words, which I find pretty annoying, but I guess then it's easy for them to convert the game for other languages.

UNIT 13

page 162, Listening: multiple choice (Part 1), Exercise 2

1

Interviewer: So Jill, tell us about the teacher who had the greatest influence on you.

Jill: Well, that's easy – Miss Cope. I was nine, rebellious and insecure when I went into her class – but three terms later I was outgoing and well adjusted.

I: So what do you put that change down to?

J: Miss Cope was inspirational. She was one of those teachers whose lessons were so enjoyable you forgot you were at school and that's what made her stand out. She was also gentle and relaxed, yet had instant control. She could make a class of children listen in rapt silence to whatever she was saying without having to shout, because it was always worth listening to. Her voice was soothing and, although she didn't have children herself, she was motherly. She seemed to be able to put up with a lot from people. I mean, even one boy who was vile to her and really pushed things to the limits. I remember her once walking out of the room, she was obviously counting to ten, before coming back in, all smiles, and starting again as if nothing had happened. I mean, even at that age I admired that – it was a life lesson in coping in itself.

2

Woman: I really thought that my graduation show looked really cool 'cos my work is very visually arresting. I love creating pieces with really vibrant plastics, you know everything from big earrings to colourful angular necklaces. But despite the fact that the show seemed to be well received, I didn't sell any of my designs afterwards, so graduating was a shock to the system. I just wondered, you know, 'What do I do now?'

Man: I know what you mean – I guess it's a good course, but you'd think they'd organise some work experience or something, you know, as well as looking good on the CV, it might open a few doors.

W: Actually, we were supposed to set that up ourselves, but I just didn't get round to it. Anyway, my tutor did help me to get a few of my designs accepted by a shop specialising in costume pieces, which was at least something – it's just that it didn't lead anywhere – so it's the right contacts they don't seem to have.

M: No well that's the main thing on these courses really.

3

Presenter: And now over to Dr Ashby, who's going to talk about the effect of computers on children's communication skills.

Dr Ashby: Thank you. You often hear it, don't you, from educationalists especially, that hours spent looking at a computer screen, playing games, surfing the net, whatever, makes kids individualistic and self-centred; that they shun social interaction and community involvement as a result. Well research into the use of chat rooms does nothing to confirm any of these fears, quite the opposite actually. Kids, it seems, use chat rooms to develop their identities and to meet others – in other words it's absolutely about developing an identity – how I fit into society, who am I, who are the cool teachers at school, and so on.

I'm reminded in all this of all the hype we had over a decade ago from just the same group of people; telling us that television, and later mobile phones, meant that kids wouldn't need to learn to read and write any more – that we were entering a new illiterate age. There was never much evidence for any of this of course and what happens? Within a few years, we have email, text messaging and all sorts of new possibilities for communicating with the written language – and kids are the first to get to grips with it!

Presenter: Thank you Dr Ashby, let me stop you there because I can see there are already one or two people who'd like to comment on …

UNIT 14

page 169, Use of English: open cloze (Part 2), Exercise 5.3

This is just to say
I have eaten
The plums
That were in
The icebox
And which
You were probably
Saving
For breakfast
Forgive me
They were delicious
So sweet
And so cold

UNIT 14

page 171, Listening: multiple choice (Part 1), Exercise 2

1

Tom: Are you running out of space to keep the press cuttings and photographs of your favourite celebrity? Spending too much time in Internet chatrooms talking about the star's latest outfit? Or are you considering plastic surgery to make yourself look more like him or her? Then you could be in the grip of Celebrity Worship Syndrome or CWS. Sally what is CWS exactly?

Sal: CWS is a term coined by psychologists to describe the increasing numbers of people who are obsessed with celebrities. They've developed a test for CWS, which identifies three levels of severity: most people simply read about their favourite celebrities as part of regular entertainment; the middle level involves discussing the star with fellow fans, while the small minority suffering from grade-three CWS exhibit more compulsive behaviour. It's only this last group who maybe have a problem, obsessing over details of a star's life and strongly identifying with their idol. The latest manifestation of this seems to be fiction on the Internet; where people write stories about their favourite celebrities. Anything can happen – Charlotte Church can save the world from global terrorism, or David Bowie can be projected into the future in a time machine and so on.

2

Presenter: Our next speaker is Derek Palmerston, the well-known educationalist, who's going to talk about homework. Derek …

Derek: Thanks. Now, let me describe a familiar scene to you. A teenager is sprawling in an armchair, listening to music while doing his homework. A well-meaning parent is looking on, getting ready to say something. It sounds like the perfect setting for a family row, doesn't it?
But let's stop for a moment. Perhaps, instead of taking the heavy-handed line of ordering him to his bedroom to get on with it quietly, maybe the parent in question should let him do the homework the way he wants. The resulting essay might just be his best ever.
Because research has shown that 20 percent of youngsters work best with background music on, 10 percent do better if they can get up and walk around the room every ten minutes and 80 percent can concentrate better if they can fiddle with a small object at the same time. So, are we guilty of trying to impose our own learning styles on youngsters? Would parents do better to keep the comments and advice to themselves, or do kids need nagging? Let's look at some of the pros and cons of each approach …

3

Woman: So did you go to the cinema on Saturday?

Man: Yeah I did actually, and I was really looking forward to the film. I mean I've been a fan of the director for years. But it was a bit of a let-down really. I mean I'm glad I went, but I don't know that I'd recommend it to you.

W: Oh it's OK, I've already seen it actually – on Sunday. I've never seen his other films, but I thought it was really well directed – especially the battle scenes – I mean they were brilliant.

M: Well, OK I'll grant you that – but scenes like that are not what he's known for – it's usually the depth of the characterisation that really makes these films and I just felt that wasn't there on this occasion.

W: Oh right. Just goes to show how we all come to these things differently, doesn't it? Anyway, I really liked the main actor, he gave a brilliant performance. What was his name?

M: I couldn't tell you.

UNIT 14

page 171, Vocabulary 2: synonyms, Exercise 3

She slammed the phone down and sat there, her heart pounding. She was shaking from head to foot and she couldn't think straight – her only instinct was to get out of the room and into a place of safety. But where could she go? Suddenly there was a loud bang on the door as someone burst in. She screamed – but even as he moved towards her she could see Carlo behind him, with three policemen. They grabbed him, and she realised she was safe.

Unit 1 test

1 Choose the most suitable verb forms to complete the following dialogues. (10 points)

A 'Hello! I (1) *didn't see / haven't seen / haven't been seeing you* for a long time! What (2) *did you do / have you done / have you been doing* since we last met?' 'Oh, working as usual.'

B The traffic was very heavy and when they (3) *arrived / have arrived / had arrived*, the concert (4) *already started / have already started / had already started*.

C I only realised towards the end of the concert that the singers (5) *didn't sing / weren't singing / haven't been singing*. They (6) *mimed / have been miming / had been miming* to a recording all the time.

D 'You and the band seem to have a very hectic schedule planned for the summer.' 'Yes indeed. By the end of the season we (7) *will perform / will have performed / will have been performing* at over 50 venues.'

E 'Has the band been together long?' 'Oh yes. By the end of this month, we (8) *will play / will have played / will be playing* together for nine years.'

F It is certainly true that the Internet (9) *had / has had / has been having* a profound effect on the way people buy music over the past 15 years, as so much can now be downloaded from websites. But I remain convinced that most of the big recording companies (10) *will survive / will have survived / will have been surviving*.

2 The following sentences contain mistakes in the part of speech used. Find each mistake and correct the word form. Some of the sentences contain **two** mistakes and some only **one**. (12 points)

1 The reviewers were not very compliment about his perform. ....
2 The inspire for Mendelssohn's Scottish Symphony was his visit to the Highlands in 1829. ....
3 To make a career as a profession musician, you need determination and dedicate. ....
4 His occasional music has declined in popular but he is still hailed as a master of orchestration. ....
5 There was thunder applause as the group made their appear on stage. ....
6 The composer's most create period was in his early 20s, when he wrote five symphonies, all of which were immediately success. ....
7 Franz Liszt was arguable the finest pianist that the world has ever known. ....
8 His parents were both musical and they gave him plenty of encourage when he was young. ....

3 Complete the second sentence so that it has a similar meaning to the first sentence, using the word given. Do not change the word given. You must use between three and six words, including the given word. (5 points)

1 The weakest part of the show was definitely the choreography. **greatest**
The .... was definitely the choreography.

2 His accomplishment as a young violinist is wonderful. **wonderfully**
He is .... young violinist.

3 Large music companies are generally reluctant to try out unknown artists. **general**
There is a .... large companies to try out unknown artists.

4 The radio controller's immediate reaction was to sack the two offending DJs. **sacking**
The radio controller .... the two offending DJs.

5 Beethoven was probably the most innovative composer of the age. **more**
Beethoven was probably responsible for .... any other composer of his age.

4 For the following questions think of **one** word only which can be used appropriately in all three sentences. (5 points)

1 The outdoor concert was a disaster because the lead singer did not turn up and on .... of that, it poured with rain.
I do not think supporting the arts is a .... priority for this government.
I heard him in the shower, singing at the .... of his voice.

2 As well as performing their own material, the band have also done some .... versions.
I didn't expect his lecture to .... so much material.
We need to charge at least £20 a ticket or we won't .... our costs.

3 It's important to .... your talk at the right level and not make it too technical.
The busker's usual .... is over there by the town hall.
The football match was disrupted when fans invaded the ....

4 I am amazed that James is auditioning for the choir because he can't sing a ....
She made a mental .... of the date of the next concert.
I don't want to act as a censor but I would sound a .... of caution about some of these lyrics.

5 In the 1970s, Abba were at the .... of their career.
Sales of *Candle in the Wind* reached a .... in the months following Diana's death.
November and December is probably the .... period for our business.

5 Choose the most suitable alternative to complete the sentences. (10 points)

1 We would like to pay .... to all the musicians who made this wonderful concert possible.
a) praise **b)** reward **c)** tribute **d)** thanks

2 When I was young, I wanted to take .... the flute. **a)** on **b)** up **c)** to **d)** out

3 You can buy fake designer clothes much more cheaply than the .... article.
a) exact **b)** real **c)** true **d)** genuine

4 His rendition of the famous 1970s classic received a rare .... ovation from the crowd.
a) standing **b)** stand **c)** stand-up **d)** standing-up

5 They did not advertise their product at first but just relied on ..... of mouth.
a) speech **b)** word **c)** news **d)** sound

6 They only had two weeks to .... the song before the recording session.
a) repeat **b)** recite **c)** rehearse **d)** resume

7 They are staging a musical production of the TV show, starring two members of the original ....
a) cast **b)** troup **c)** set **d)** script

8 The lead singer was dreadful as he sang out of .... for most of the time.
a) note **b)** pitch **c)** melody **d)** tune

9 The audience fell silent as the .... stood in front of the orchestra and raised his baton.
a) leader **b)** director **c)** manager **d)** conductor

10 I can't remember the words but it is a really .... tune.
a) gripping **b)** catchy **c)** holding **d)** sticky

6 Some of the following sentences contain a wrong preposition. Find the wrong prepositions and correct them. If all the prepositions in a sentence are correct, put a tick at the end. (8 points)

1 I am writing to complain for the late arrival of the tickets. ....
2 We would like to compliment you on your brilliant performance. ....
3 He decided to give up his job and concentrate at a career in music. ....
4 She is crazy for Latin American music, especially swing. ....
5 Their latest album is the best they have produced by far. ....
6 The concert could not continue owing from faulty sound equipment. ....
7 He has never shown any interest in learning a musical instrument. ....
8 A string quartet usually consists in two violins, a viola and a cello. ....

Unit 2 test

1 Complete the following text with *a/an*, *the* or (–) when no article is needed. (15 points)

Most people would agree that (1) .... credit cards can be invaluable in (2) .... emergency or when travelling abroad. But (3) .... main problem with them is that some people overuse them and eventually end up deep in (4) .... debt. If you find yourself in this situation, it is important to plan how you are going to pay it off. Firstly, you need to make (5) .... real effort to pay more than (6) .... minimum required each month. Examine your daily routine; there must be (7) .... few sacrifices you can make. Take (8) .... sandwiches to (9) .... work, for example, instead of eating out. Look at your debts carefully too and make sure that you pay off (10) .... one with (11) .... highest interest rate first. And in future? If you think that there is (12) .... chance that your spending will get out of (13) .... control again, perhaps you should cut (14) .... credit cards up. After all, living without them is probably better than living in (15) .... fear of those bills every month.

2 Complete the sentences using a relative pronoun from the box. Use each pronoun **once** only. (5 points)

that	which	who	where	whose

1 Duncan Reed, ... appeared in the famous jeans commercial, is now hosting a quiz show.
2 Did you speak to the lady .... handbag was stolen?
3 The question .... you must ask before you make any purchase is 'Do I really need it?'
4 Starbucks, ... was launched in the 1970s, has become one of the most successful chains of coffee shops.
5 The building .... the event is being held is actually a disused power station.

3 Correct the punctuation mistake in each of the following sentences. You can either add or delete a punctuation mark. (5 points)

1 He reported that, the charity had received over £10,000 in donations.
2 Mrs Howard's necklace, which had been in her family for generations was put up for auction.
3 She fell in love with the house when she saw it's beautiful gardens.
4 The money, that I put aside for the rent has disappeared.
5 People who buy cheap second-hand clothes should ask themselves how long they want them to last?

4 For the following questions think of **one** word only which can be used appropriately in all three sentences. (5 points)

1 It is .... to assume that the shares will increase in value in the long term.
   She did not know what to buy for his birthday but thought a book token would be a .... bet.
   I can't give you any money now because it's all locked away in the ....
2 She was born in the .... north of Scotland.
   Unfortunately, profits fell .... short of what was expected.
   Her explanation sounded rather ....-fetched and I don't think I believe it.
3 A train crashed into a car on the ....-crossing last night.
   He's a very ....-headed person and never panics in a crisis.
   The book only analyses the data on a very basic ....

4 After winning the quiz, the team were in .... spirits.
The minister made a ....-profile visit to the north of the country.
It's .... time you settled down to some serious studying.

5 The plans to build a new airport do not take into .... its effect on the environment.
On no .... are you to leave the office door unlocked.
I keep most of my savings in a deposit ....

5 Complete the following sentences using the correct form of the word in bold. (10 points)

1 It is very important to take steps to reduce energy .... **CONSUME**
2 The minister was strongly criticised for showing .... to his party. **LOYAL**
3 They have a system which allows you to take back .... goods to be recycled. **WANT**
4 Some patients may develop an emotional .... to their nurse or counsellor. **ATTACH**
5 The live music and the costumes made the festival a very .... occasion. **MEMORY**
6 Examiners can use their .... over whether to award a pass to unfinished scripts. **DISCERN**
7 I would like to work freelance but I am worried about the .... **SECURE**
8 With the increase in our mortgage rate, we will be forced to .... **ECONOMY**
9 It was a nice bright room, but the furnishings were very .... **ATTRACTIVE**
10 The current political situation is unstable and .... very dangerous. **POTENT**

6 Choose the most suitable alternative to complete the sentences. (10 points)

1 Buying a flat to rent out was a very .... investment.
a) firm b) rich c) sound d) sturdy
2 You should put some money aside for a .... day.
a) wet b) rainy c) cold d) dull
3 The company made a good profit in its first year but now it's over 10 thousand pounds in the ....
a) red b) black c) pink d) green
4 I think we need to .... a different approach to the problem.
a) mind b) raise c) nourish d) adopt
5 If you .... around on some of the market stalls, you can find some real bargains.
a) ransack b) browse c) rummage d) surf
6 It is important to plan your spending and .... yourself a budget.
a) set b) lay c) place d) establish
7 By the end of their studies, many students were .... in debt.
a) highly b) widely c) heavily d) intensely
8 In total, the repairs will cost somewhere in the .... of £3000.
a) region b) field c) area d) zone
9 She won't be pleased by your decision but you'll just have to bite the .... and tell her.
a) nail b) bullet c) needle d) key
10 He wrote me a cheque but, to my annoyance, last week I found that it had ....
a) jumped b) sprung c) bounced d) leapt

Unit 3 test

1 Complete the following text with the correct modal expression from the three alternatives. If there are no alternatives, put in one verb from the box below in either the gerund or the infinitive form. (18 points)

put spend demand work limit look move draw forgive

FORGIVENESS

The possibility of revenge was part of the system of justice in many ancient societies. It survives today in proverbs such as 'an eye for an eye' or 'sauce for the goose is sauce for the gander'. Indeed, these principles were an attempt (1) .... the effect of revenge by laying down that the avenger's action (2) .... exactly the same amount of harm as the original action. Otherwise, the need to take revenge (3) .... to a series of ever more harmful acts, with no obvious end in sight.

However, there is an alternative to the practice of taking revenge and that is to stop (4) .... that the wrongdoers (5) .... punished; in other words, to forgive them. Forgiveness has often been associated with religious teachings but in fact you (6) .... religious to practise it. Psychological studies show that bearing a grudge is typically a source of stress and unhappiness, while forgiving someone allows the wronged person (7) .... the event behind them. However serious the wrongdoing, there is a need in the end (8) .... a line in the sand, otherwise we risk (9) .... the rest of our lives eaten up with bitterness and anger.

As an extreme example of the power of forgiveness, we (10) .... the story of Amy Biehl. Amy was murdered in South Africa while she was working to assist the anti-apartheid movement. After some years of extreme grief, her parents decided (11) .... to South Africa to continue her work, and there, they met her two killers. These two young men applied (12) .... for the foundation that Amy's parents had established in her name and in time they became close friends with the parents, eventually even calling Amy's mother 'mum'. The fact that Mr and Mrs Biehl managed (13) .... the death of their daughter (14) .... incredible to some people. Yet, while it is true that Amy's parents (15) .... exceptional people to have practised this degree of forgiveness, we should also reflect on the alternative. Whatever her parents did, they (16) .... Amy back. They (17) .... the rest of their lives feeling only grief and anger. But instead, forgiving enabled them (18) .... to the future with hope, as part of the South African movement towards reconciliation. Even from their own point of view, they were better off than if they had not forgiven.

2 a) had to give b) must give c) must have given
3 a) must have led b) can lead c) could have led
5 a) should be b) might be c) could be
6 a) mustn't be b) don't have to be c) shouldn't be
10 a) might take b) have to take c) may have taken
14 a) may well seem b) must well seem c) should well seem
15 a) must have been b) should have been c) ought to have been
16 a) might not have brought b) could not have brought c) may not have brought
17 a) may have spent b) must have spent c) could have spent

2 Complete the second sentence so that it has a similar meaning to the first sentence, using the word given. Do not change the word given. You must use between three and six words, including the word given. (10 points)

1 It would be a good idea to give the candidates a compulsory personality test. **take**
It would be a good idea if the candidates .... a personality test.

2 I think it is better if you don't contact him for the moment. **to**
I don't think .... him for the moment.

3 Attending the seminar is optional for first-year students. **have**
First-year students .... attend the seminar.

4 If you enjoyed his last novel, how about trying some of the earlier ones? **like**
If you enjoyed his last novel, you .... try some of his earlier ones.

5 It was a bad idea for him to lie to his parents like that. **should**
He .... than to lie to his parents like that.

6 He continued to take the medicine even after the symptoms had disappeared. **on**
He .... the medicine even after the symptoms had disappeared.

7 These areas of the brain are used in recognising facial expressions. **allow**
These areas of the brain .... facial expressions.

8 I don't think that parents should argue in front of their children. **avoid**
I think that parents .... in front of their children.

9 I have no intention of listening to your excuses. **prepared**
I .... to your excuses.

10 You can't make him go to the party if he doesn't want to. **force**
You can't .... to the party if he doesn't want to.

3 Match each personality adjective in the box with the correct description. (10 points)

arrogant quick-tempered high-spirited taciturn vindictive conscientious gullible inquisitive self-conscious trustworthy

1 He takes his work very seriously and always does things properly. ....
2 He doesn't talk much and can seem rather unfriendly. ....
3 He can become very angry over quite trivial things. ....
4 He has too high an opinion of himself and doesn't consider others. ....
5 He is easily embarrassed and often worried about how he appears to other people. ....
6 He's always trying to find out about other people's lives and what they are doing. ....
7 You can easily trick him as he tends to believe whatever you tell him. ....
8 He's very honest and reliable. ....
9 He doesn't forget if someone upsets him and always tries to get back at them. ....
10 He's very lively and energetic. ....

4 Choose the most suitable alternative to complete the sentences. (12 points)

1 That student seems very .... and hardly says a word in class.
a) outgoing b) underhand c) withdrawn d) downbeat

2 To all the people who contributed, we would like to express our .... thanks.
a) hearty b) heartfelt c) heartened d) heartrending

3 The local authority .... down over their proposal to close the hospital.
a) moved b) turned c) stepped d) backed

4 I didn't realise he felt so strongly so I was surprised at his .... of anger.
a) outcry b) outburst c) outlet d) output

5 You'll certainly .... the crowd with that bright jacket!
a) stand out from b) stand up to c) stand in for d) stand away from

6 Peter cheated me out of the money but I'll get .... with him somehow.
a) level b) even c) equal d) smooth

7 She's very sensitive about her looks so don't .... fun of her.
a) take b) make c) do d) carry

8 The meeting finished early so they made a .... decision to go out for lunch.
a) crack b) snap c) smash d) clap

9 His punishment should .... as a warning for those considering similar offences.
a) work b) show c) use d) serve

10 I really regret .... school at the age of 16.
a) to leave b) leave c) leaving d) to have left

11 The study .... out to prove that certain personality types are more prone to heart disease.
a) sets b) carries c) takes b) gives

12 She .... the impression of being very nervous in the interview.
a) made b) gave c) did d) put

Unit 4 test

1 Complete the second sentence so that it has a similar meaning to the first sentence, using the word given. Do not change the word given. You must use between three and six words, including the word given. (10 points)

1 The patient only survived because he was given the new antibiotic. **had**
The patient would have died if he .... the new antibiotic.

2 A great many misunderstandings are caused by people sending unclear emails. **were**
If people .... emails, there would not be so many misunderstandings.

3 Please feel free to ring our helpline if you require further assistance. **should**
Please feel free to ring our helpline .... further assistance.

4 The crash was caused by a failure to carry out checks on the signals. **had**
The crash would not have occurred if .... on the signals.

5 If I met the inventor, I would ask for his autograph. **meet**
Were ...., I would ask for his autograph.

6 How would you react to being cared for by a robot nurse? **after**
Supposing ...., by a robot nurse, how would you react?

7 You can't access the programme unless you know the right password. **happen**
You can only access the programme .... the right password.

8 I am sorry I bought this car because it is so unreliable. **realised**
Had ...., I would not have bought this car.

9 Giving parents the possibility of choosing the sex of their child could result in a population imbalance. **given**
If .... of choosing the sex of their child, this could result in a population imbalance.

10 Our products look out of date because we do not invest enough in research and development. **invest**
If .... in research and development, our products would not look so out of date.

2 Correct the following sentences. In each case, you must either **add** one word, or **remove** one word. (10 points)

1 We do not know enough about what the long-term effects on genetically modified crops will be if we will start cultivating them on a large scale. ....
2 If we had known about the side effects, we would not put the drug on the market. ....
3 Unless we are not responsive to our customers' needs, our market share will continue to fall. ....
4 If the Wright brothers had not persisted in their experiments, aeroplanes might never have invented. ....
5 If you happen see James, tell him to call me on the mobile. ....
6 Had the CCTV working, we could have obtained pictures of the thieves. ....
7 The machinery should last a long time provided with it is serviced regularly. ....
8 If we were concentrate on developing renewable energy sources, we would be much closer to solving the problem of climate change. ....
9 If I had have known that the software was so expensive, I would not have ordered it. ....
10 If it had not been the discovery of DNA, a great many serious crimes would have remained unsolved. ....

3 In each case, cross out the noun which does **not** collocate with the given verb. (10 points)

1 catch a cold, fire, control of something, sight of something
2 give someone a chance, a question, someone a hand, a shout
3 lose contact, issue, your way, your temper
4 keep a promise, guard, charge of something, someone company

5 hold quiet, your breath, the record, a meeting
6 make an effort, a favour, your living, progress
7 have a party, a rest, a breath, an argument
8 set a picture, the alarm, an example, the scene
9 pull a muscle, someone's leg, a face, your tongue
10 draw a difference, the curtains, a sketch, a big audience

4 Complete each sentence with the correct word from the box to make a fixed phrase. There are more options than you need. (5 points)

matter sense shred pack twist move rule

1 He may well claim he knows best, but that's really only a .... of opinion.
2 The antibiotic was only discovered by a strange .... of fate.
3 As a .... of thumb, you should leave about 20 minutes for the chemical to work.
4 They claim to have cloned several human beings but I think it is probably all a .... of lies.
5 There is not really a .... of evidence to suggest that computer games are a cause of violent behaviour.

5 For the following questions think of **one** word which can be used appropriately in all three sentences. (5 points)

1 I tried to understand the instructions but they did not make ....
Having a close family is important because it gives a child a .... of security.
It isn't difficult to understand if you use your common ....
2 I think it's lack of communication which is at the .... of the problem.
I'd need a calculator to work out the square .... of that.
The main ingredients are tomatoes and a mixture of different .... vegetables.
3 Some scientific discoveries are made by pure chance and penicillin is a .... in point.
The report makes out a strong .... for investing more in research and development.
Pack your .... the night before to save time in the morning!
4 It has a reputation as a .... chip company and should be a very safe investment.
That decision came completely out of the ....; I wasn't expecting it at all.
Her hands were .... with cold when she came in from playing in the snow.
5 Unfortunately, politics cannot be reduced to an exact ...., and the unexpected can often happen.
I enjoy reading novels especially detective stories and .... fiction.
They are taking on a new lecturer in the social .... department.

6 Choose the most suitable alternative to complete the sentences. (10 points)

1 The government are opening a new science park but it's the taxpayer who will .... the bill.
a) head b) back c) foot d) finger
2 I was .... disappointed that no one wanted to invest in our product. a) bitterly b) sourly c) heavily d) sharply
3 I advise you to .... a low profile until the investigation is over. a) take b) make c) keep d) set
4 Some people feel that it is wrong to .... experiments on animals. a) lead b) conduct c) direct d) guide
5 We are .... grateful to all those people who agreed to take part in the experiment.
a) highly b) deeply c) strongly d) firmly
6 The research .... up a number of interesting issues. a) pulled b) held c) drove d) threw
7 If some experts are correct, the technological revolution is only in its ....
a) childhood b) infancy c) youth d) nativity
8 The word *tachyon* was .... to refer to a particle which supposedly travels faster than light.
a) minted b) cast c) coined d) moulded
9 They won the game because they were talented; it was no .... a) flick b) fluke c) fling d) flip
10 The best way to remember the procedure is to break it .... into three stages. a) out b) over c) away d) down

Progress test 1 (Units 1–5)

1 Complete the following sentences using the correct form of the word in bold. You will need to add suffixes in each case. (15 points)

1 He made a very good ………….. in the job interview. **IMPRESS**
2 House prices in that region have become ………….. high. **ASTRONOMY**
3 A professional sportsperson's life may look ………….. but it involves a lot of hard work. **GLAMOUR**
4 There was a wonderful ………….. of products on display at the trade fair. **VARY**
5 I found the story of how he became an entrepreneur quite ………….. . **INTRIGUE**
6 She makes a fair number of enemies because of her ………….. style. **CONFRONT**
7 He had set his heart on winning so he was really disappointed to be just the ………….. up. **RUN**
8 In general, I think the management team would be ………….. of the idea. **SUPPORT**
9 ………….., the team managed to win three-nil, even though they had two players missing. **AMAZE**
10 He seems pleasant to talk to but I don't really believe he is ………….. . **TRUST**
11 Despite the difficulties, caring for elderly people can be a very ………….. job. **REWARD**
12 I am afraid I ………….. deleted one of the files on your computer. **ACCIDENT**
13 I refuse to take ………….. for other people's mistakes. **RESPONSIBLE**
14 There seems to be a growing public ………….. in environmental projects. **INVOLVE**
15 The exercises will help to keep you supple and develop your ………….. . **FLEXIBLE**

2 Each of the following sentences contains one word which is incorrectly used. Cross it out and write a correct word. (10 points)

1 I'm really sorry but I've just been crashing into your car. ………………………………
2 You ought be strong enough to deal with the problem on your own. ………………………………
3 The Royal Arms Hotel, that was refurbished last year, is now definitely the best in town. ………………………………
4 Take some waterproof clothes with you just if you need them. ………………………………
5 The hospital which I used to work is in North London. ………………………………
6 By the time I arrived at the party, everyone went home. ………………………………
7 If you want to pass the exam, you would have to improve your essay writing. ………………………………
8 That was an obvious attempt making me look stupid in public. ………………………………
9 Supposing I only came to the second part of the meeting, will that be a problem? ………………………………
10 If I were report you to the authorities, what would you do? ………………………………

3 There are ten unnecessary articles (*a* or *the*) in the following text. Find them and cross them out. (10 points)

ADVERTISING THROUGH SMELL

We are all used to advertisers using the sight and sound to grab our attention. Colourful logos are in an evidence in our cities at every turn and TV commercials assail us with slogans and jingles whenever there is a break in the programmes. But the most of us are not so aware that we can also be targeted through our sense of the smell. The technique is not a new one. Supermarkets deliberately reheat their bread on the premises in the hope that the aroma will tempt shoppers to go to the bakery section on the impulse. In the same way, many coffee shops like to grind a coffee at the bar so that passers-by will be drawn in by the delicious smell. Now, however, a new device has been developed by a Japanese inventor, called an air cannon. This can single out a particular consumer and shoot a specified smell directly up their nose. With an equipment like this, retail outlets are no longer confined to one or two long-lasting smells. The person by the fruit counter could get a whiff of citrus while another shopper looking for a cleaning stuff could be targeted with a smell of the beeswax or pine detergent. The result, no doubt, will be a greater number of purchases made without a due consideration and, of course, more money for the stores.

4 Complete each of the following sentences with a word formed from one of the alternatives in box A and a prefix in box B. There are more prefixes than you need. (10 points)

A

logical	developed	war	reversible	biography
climax	performed	understood	spent	obedient

B

un-	post-	ir-	mono-	mis-	under-	over-	auto-
dis-	ir-	fore-	in-	il-	anti-	semi-	out-	im-

1 After all the excitement created by the media, the actual match turned out to be a slight ....
2 She was the player with the lowest score, but that was because she .... the rules.
3 He's a very young player but he definitely .... the rest of the team.
4 People are often more disappointed if they come second than if they come last, ... as that may seem.
5 Harry Taylor, the famous businessman, published his .... last year.
6 That part of the city was very .... until they built the new shopping centre there.
7 She's very .... and takes no notice of my instructions at all.
8 It is feared that the patient's health has gone into .... decline.
9 Despite their reassurances that the cost would be kept down, by the time the new sports complex was finished, they had .... their budget by £4m.
10 That big housing complex was built in the .... years.

5 In the following sentences, **two** of the alternatives are correct and **one** is not possible. Cross out the incorrect alternative. (10 points)

1 It is *extremely / very / absolutely* important to do some warming up exercises before you start to work out.
2 Last year my friend took me on a climbing trip to the Alps. It was a / an *really / absolutely / very* wonderful experience.
3 The rules of the competition are *absolutely / utterly / crystal* clear. Any athlete who fails the drugs test will be disqualified.
4 Well fitting trainers are *absolutely / very / really* essential for long-distance running.
5 I was caught in the rain without an umbrella this morning. I got absolutely *wet / soaked / drenched*.
6 The tennis final was *really / very / absolutely* exciting. I couldn't take my eyes off the screen.
7 I could not find a seat on the train as it was *absolutely / extremely / completely* packed at that time in the morning.
8 She was extremely *disappointed / devastated / upset* not to have even been shortlisted for the job.
9 He must be one of the greatest footballers ever. His control of the ball is absolutely *skilful / masterly / second to none*.
10 I felt *totally / terribly / completely* exhausted after my game of squash.

6 Complete the second sentence so that it has a similar meaning to the first sentence, using the word given. Do not change the word given. You must use between three and six words, including the given word. (15 points)

1 Your questionnaire can remain anonymous if you prefer. **have**
You .... your name on the questionnaire.
2 Their equipment is subjected to more rigorous testing than any other manufacturer's. **tested**
Their equipment .... than any other manufacturer's.
3 In the past, she exercised more often than she does now. **as**
She does not exercise.. to.
4 It is commonly assumed that good teachers have to be extrovert. **common**
There is .... good teachers have to be extrovert.
5 He'll probably forget all about the appointment. **chances**
The .... all about the appointment.
6 Membership of the gym has fallen dramatically this year. **far**
The gym has .... than last year.
7 Should you see Keith this afternoon, give him my regards. **happen**
Give Keith my regards .... this afternoon.
8 It's a great shame that I didn't ask her for her mobile number. **regret**
I really .... for her mobile number.
9 The exercises were much easier than I thought they would be. **nearly**
The exercises were ....
I thought they would be.
10 If you've forgotten the combination, you can't open the safe. **remember**
Unless ...., you can't open the safe.
11 Support for the Green Party has risen dramatically. **rise**
There has been .... for the Green Party.
12 We arrived too late to hear the president's speech. **could**
If we had arrived earlier, we ....
the president's speech.
13 Our high intelligence means that we are able to foresee the consequences of our actions. **allows**
Our high intelligence .... the consequences of our actions.

14 I paid too much for the necklace because I didn't know that they were not real pearls. **known**

Had .... imitation, I would not have paid so much for the necklace.

15 These trainers are much better than any other brand. **superior**

These trainers .... any other brand.

7 Choose the most suitable alternative to complete the sentences. (20 points)

1 They went on to win the championship, against all the ....

a) chances b) odds c) fortunes d) stakes

2 The team suffered a major .... when their best player retired, due to ill health.

a) setback b) throwback c) comeback d) fallback

3 They went on a shopping .... together that weekend.

a) rave b) jaunt c) spree d) function

4 The tennis champion easily beat his .... in three sets.

a) competitor b) opponent c) contestant d) enemy

5 He's a bit of a/an .... To my knowledge, he has never won any games or even come second.

a) also-ran b) no show c) has been d) high flyer

6 We can't book the holiday until we know whether we will be allowed time off, so don't jump the ....

a) pistol b) gun c) shot d) rifle

7 He finally .... his ambition to have his own television show.

a) reached b) gained c) won d) fulfilled

8 Since the club opened last year, the enthusiasm for keeping fit has really taken ....

a) on b) off c) up d) to

9 I never touch the money in that fund as it's my ....

a) egghead b) good egg c) nest egg d) eggshell

10 It is impossible to predict the .... of the election as the vote seems split about 50-50.

a) outlook b) outgoings c) outcome d) outlay

11 James seems to be the only suitable candidate for the job now that Maria is out of the ....

a) running b) racing c) jumping d) driving

12 He's not a complete .... when it comes to skiing as he took some lessons in the past.

a) novel b) novice c) innovator d) novelty

13 They were married last year but split .... after only six months.

a) off b) away c) out d) up

14 We have a long-........... agreement with our suppliers.

a) suffering b) standing c) winded d) lived

15 He was driven by an .... desire to make the company a success.

a) overcoming b) overjoyed c) overwhelming d) overstretched

16 I can't lend you anything at the moment as I'm really .... up.

a) tough b) hard c) firm d) solid

17 He's amusing to talk to because he is so quick ....

a) brained b) minded c) witted d) headed

18 You need to eat three good meals a day to .... up your strength.

a) raise b) make c) form d) build

19 I think the system will work well eventually but it still needs a few ....

a) twists b) tweaks c) twirls d) twitches

20 I wasn't expecting the news at all; it came out of the ....

a) black b) red c) green d) blue

8 Decide if the following statements about types of text are true or false. (10 points)

1 The first paragraph of a formal letter usually gives the reason for writing. ....

2 'All the best' is a suitable ending for a formal letter. ....

3 If you begin a formal letter 'Dear Mr Smith', you should finish 'Yours faithfully'. ....

4 The final paragraph of a formal letter often contains a request or suggestion. ....

5 An information sheet can use bullet points. ....

6 '*Dos* and *don'ts*' is a common subheading in information sheets. ....

7 An article often uses bullet points. ....

8 An article does not need to be written in paragraphs. ....

9 An article should discuss both sides of the question before giving your opinion. ....

10 A reference can begin with a rhetorical question to catch the reader's attention. ....

Unit 6 test

1 Complete the second sentence so that it has a similar meaning to the first sentence, using the word given. Do not change the word given. You must use between three and six words, including the given word. (10 points)

1 I don't want you to send a wedding invitation to your ex-boyfriend. **invite**
I'd rather .... your ex-boyfriend to the wedding.

2 I should have taken your advice instead of thinking that I knew best. **followed**
I wish .... your advice instead of thinking that I knew best.

3 It's a pity that I have forgotten her phone number. **remember**
I wish .... her phone number.

4 Annalisa and I can't get married because her father won't agree to it. **consent**
If only Annalisa's father ...., we could get married.

5 It's a pity my house is so far away from my daughter's. **lived**
I wish .... to my daughter's house.

6 I don't like it when you stare at me like that. **stop**
I wish .... at me like that.

7 You really should realise that money does not grow on trees. **time**
It's about .... that money does not grow on trees.

8 I'm sorry I didn't allow you to help me now, as it was more difficult than I thought. **let**
I wish .... help me now, as it was more difficult than I thought.

9 I don't want you to keep anything from me, even if it's bad news. **told**
I'd rather .... even if it's bad news.

10 It was a real mistake to argue on our wedding anniversary. **argument**
If only .... on our wedding anniversary.

2 Complete each response by inserting a substitute word. (10 points)

1 'Is your son happy in his new job?' 'Well, I don't think ....'

2 'Did you enjoy the film?' 'No, I didn't really like ....'

3 'Excuse me, you've just trodden on my foot.' 'Sorry, I didn't mean ....'

4 'There was a fight at the disco last night. I was really scared.'
'Well, I wouldn't go .... again if I were you.'

5 'Is Aunt Marian spending Christmas with us this year?' 'Yes, I expect ....'

6 'Did he give a reason for dropping out of university?'
'Yes, he gave two or three reasons, but not very good ....'

7 'Was he sorry for speaking to you like that?' 'Well he certainly didn't say ....'

8 'The play starts at 7, so we could meet for a drink at 6.30.' 'OK, I'll see you ....'

9 'Did you get married just because it was what your parents wanted?'
'Certainly not. I chose ....'

10 'Do you think couples should stay together just because of the children?'
'Hmm. That's a difficult .... to answer.'

3 Complete each response with an auxiliary verb. Some gaps may need more than one verb. (6 points)

1 'Do you think he is jealous of his elder brother?' 'Yes, I think he might well ....'
2 'Will you take the night off and go to the concert?' 'I may .... It depends how busy I am.'
3 'There was no answer when I called Jane. Maybe she's visiting her sister.' 'She could ....'
4 'Where's Robert? Has he gone home without us?' 'He can't .... I've still got the keys to his car.'
5 'Shall we go and see Peter and Karen while we're in town?' 'We could .... if you like.'
6 'Oh no! Do you think he was listening to us outside the door?' 'Well, I suppose he might ....'

4 Each of the following sentences contains a missing preposition. Insert it in the correct place. (10 points)

1 I don't know how my parents will react that idea.
2 All the family congratulated him passing his driving test.
3 My daughter has a very short attention span and can't concentrate anything for long.
4 I wish people wouldn't keep comparing me my elder sister.
5 She apologised her mother for staying out so late.
6 His wife is bedridden and depends him for absolutely everything.
7 You've done very well in your exams but I wish you would stop boasting it.
8 Rising house prices have resulted many young wives agreeing to live with their in-laws.
9 He's always complaining his son spending so much time on the telephone.
10 If you can't talk to your parents, do you have an elder brother or sister you could confide?

5 Each of the following sentences contains one word which is incorrectly spelt. Cross it out and write a correct word. (6 points)

1 She gave him some good advise on how to deal with his wayward son. ....
2 She refuses to accept money from her family as a matter of principal. ....
3 We must take care not to loose contact now you're off to university. ....
4 He crashed his father's car into a stationery vehicle. ....
5 His parents' divorce had a long-lasting affect on his ability to form relationships. ....
6 I wish your son wouldn't practice his guitar while I'm trying to work. ....

6 Choose the most suitable alternative to complete the sentences. (8 points)

1 We were a close couple at first but after a few years we seemed to .... apart.
a) float b) glide c) drift d) sail
2 Because of the rapidly changing society here, you often find that there is a big generation ....
a) space b) gap c) split d) hole
3 I am worried that my daughter may be .... by other children because of her size.
a) picked on b) picked up c) picked at d) picked off
4 He thought she was wonderful at first but the initial attraction soon ....
a) came off b) fell off c) wore off d) took off
5 Marriage is intended to be a .... commitment.
a) long-running b) long-life c) lifelike d) lifelong
6 I'm afraid Pat wrote to me this week to say that her marriage is ....
a) breaking off b) breaking up c) breaking down d) breaking away
7 I'm afraid you have to .... down the law to your son and tell him that he has to be back before midnight.
a) lay b) put c) set d) hold
8 Recently a number of new office blocks have .... up in the eastern side of the town.
a) jumped b) sprung c) risen d) leapt

Unit 7 test

1 Choose the most suitable verb forms to complete the following dialogues. (10 points)

1 'What are your plans for this evening?'
'*I'll stay / I'm going to stay / I stay* at home and write my Christmas cards. I can't put it off any longer.'
2 'Why do we have to be here so early tomorrow morning?'
'Because the photographer *comes / is coming / will come* at 9 o'clock.'
3 'Could I pick up the portrait at 5.00 tomorrow?'
'No, sorry. *I'll leave / I'll have left / I'm leaving* the studio by then. '
4 'I'm just off to have a look at that new poster exhibition.'
'That's a good idea. I think *I'll come / I'm coming / I'm going to come* with you.'
5 'Do you have any plans for July?'
'Yes, I've booked a holiday course in Florence. *I'll study / I'll be studying / I study* Renaissance art.'
6 'Do you think I should hang the picture on that wall?'
'No, put it above the fireplace. It *will be / will have been / is going to be* more of a focal point there.'
7 'Don't lean on that glass case like that! *You'll break / You'll be breaking / You're going to break* it!'
8 'By the end of next week, he *is going to finish / is finishing / will have finished* the painting.'
9 'What time *does your plane arrive / will your plane arrive / is your plane going to arrive*?'
10 'This time tomorrow, *we'll enjoy / we'll be enjoying / we're going to enjoy* the sights of Rome.'

2 Complete the second sentence so that it has a similar meaning to the first sentence, using the word given. Do not change the given word. You must use between three and six words, including the given word. (5 points)

1 He was just going to confess that the painting was a forgery. **point**
He was .... that the painting was a forgery.
2 I am not going to put that painting up for auction. **intention**
I .... that painting up for auction.
3 The first day of the Aztec exhibition is next week. **due**
The Aztec exhibition .... next week.
4 The bird flew away before I could get my camera out. **about**
I .... my camera out when the bird flew away.
5 I think she'll be shocked when she finds out how much the painting is worth. **for**
I think she .... when she finds out how much the painting is worth.

3 In the following text, put the verbs in brackets in the correct tense or form. Choose from: infinitive with *to*, infinitive without *to* and gerund. In some cases you need to add a pronoun as well. (18 points)

Stories of famous forgers have a particular fascination, especially when they manage (1) .... (deceive) the most eminent art critics of the day. The most notorious art forger of the 20th century was Han Van Meegeren, originally from the Dutch town of Deventer. As a child, he loved (2) .... (draw) and planned even then (3) .... (make) it his career. He received professional recognition at high school and went on (4) .... (study) architecture. Ten years later, he was earning his living by teaching art and painting portraits.

Van Meegeren, however, was a very conservative artist, whose works were essentially imitative of an earlier tradition. As a result, critics frequently ridiculed his work as uninspired and derivative. In the end, he decided (5) .... (revenge) himself on the art world. He set out (6) .... (paint) a forgery, an imitation of the Dutch artist Vermeer, which he would pass off as a newly discovered work. Although he had hoped (7) .... (fool) a few eminent critics, nothing could prepare him for the success of his hoax. It was eventually sold to the gallery in Rotterdam for the equivalent of 2 million dollars.

Van Meegeren had originally intended (8) .... (humiliate) the art world by confessing the truth, but after this success he changed his mind. Instead, he kept on (9) .... (produce) fakes and produced a total of six 'Vermeers' in the following years. They made him a very rich man indeed.

The truth only came out towards the end of the Second World War. When he sold one of his fake Vermeers, fatally he allowed (10).......... (fall) into the hands of the enemy. In order to avoid (11) .... (serve) a long prison sentence for collaboration with the Nazis, he admitted (12) .... (forge) the painting. At first, critics refused (13) .... (believe) him. They were only convinced after he persuaded (14) .... (let) (15) .... (paint) a new fake Vermeer in his prison cell. The resulting painting was clearly by the same hand as the others.

The beneficial effect of a case like that of Van Meegeren is that it forces (16) .... (re-examine) the way we judge art, to stop (17) .... (be) so concerned with simple authenticity and to ask ourselves what we should really be looking for in a painting. Van Meegeren may not have been a great artist but he makes (18) .... (think) harder about what we value in art and why.

4 Complete the following sentences with a phrasal verb from the box. In each case, you also need to change the tense. (5 points)

dash off	bring about	clear out	come across	snap up

1 Most of the people at the auction were antique dealers, who had set their hopes on .... some items of rare furniture.

2 Yesterday, she .... a pile of old prints while she .... the attic.

3 I can't believe you think it's a great work of art. It looks as if it .... in a matter of seconds.

4 Caravaggio's use of dramatic light effects .... a revolution in devotional painting.

5 Choose the most suitable alternative to complete the sentences. (12 points)

1 Many countries would welcome the chance to .... an event like the Olympic Games.
a) receive b) host c) accept d) guest

2 She sold the painting and bought a new flat with the ....
a) procedure b) process c) proceeds d) proceedings

3 He couldn't believe it when he saw his picture .... across the front page of the newspaper.
a) poured b) gushed c) splashed d) squirted

4 They are planning to .... a production of Puccini's Madam Butterfly.
a) put on b) put out c) put up d) put in

5 He was caught .... stationery from work and nearly lost his job.
a) burgling b) poaching c) looting d) pilfering

6 The painting was not recognised as a great work and was eventually sold for the .... sum of £100.
a) modest b) humble c) restrained d) plain

7 Many collectors are willing to pay over the .... for one of his early paintings.
a) rates b) odds c) bets d) evens

8 It is an impressive work at first but it does not really .... close examination.
a) stand up to b) stand out from c) stand up for d) stand in for

9 Unfortunately, I think there is one very serious .... in your argument. a) rip b) blemish c) split d) flaw

10 The lecturer .... a number of parallels between Blake's engravings and devotional art.
a) drew b) pulled c) tied d) linked

11 He was so .... on his work that he lost all track of time.
a) intent b) intentional c) intense d) intensive

12 It is an .... that the most talented artists often go completely unrecognised in their lifetime.
a) irregularity b) eccentricity c) anomaly d) abnormality

Unit 8 test

1 John, a company owner, is speaking to Louise, one of his sales reps, who he feels has lost her motivation. Read the extract from their conversation below, then complete the report by changing the verbs into reported speech. (8 points)

You have so much to offer this company. You contributed a lot during the first few months but now I feel you've lost your enthusiasm. The problem is if you don't bring in more business, we may not make it through next year.

OK, I'll try harder. But don't just blame the reps. We should evaluate our pricing system as well. Several of our competitors have managed to undercut us.

I pointed out that Louise (1) .... a great deal during the first few months but I (2) .... that she (3) .... her enthusiasm. I warned her that if she (4) .... in more business, we (5) .... not make it through next year.

Louise said that she (6) .... harder but that we (7) .... evaluate our pricing system as well, as several of our competitors (8) .... to undercut us.

2 Choose the most suitable alternative to complete the sentences. (10 points)

1 The consultant suggested .... our staff on a training course. **a)** us to send **b)** us sending **c)** that we should send
2 The manager accused Karen .... online shopping during work hours. **a)** to do **b)** of doing **c)** on doing
3 He insisted .... high targets for the next financial year. **a)** to set **b)** of setting **c)** on setting
4 He complained .... to work late every evening. **a)** about having **b)** of having **c)** to have
5 I tried to persuade .... a larger quantity of the product. **a)** him to buy **b)** to him to buy **c)** him of buying
6 They urged .... his lawyer immediately. **a)** him to contact **b)** on him contacting **c)** that he should contact
7 He admitted .... the truth on his application letter. **a)** to stretch **b)** stretching **c)** of stretching
8 He threatened .... if the new contract was not modified. **a)** me to resign **b)** to resign **c)** of resigning
9 He congratulated her .... winning the award. **a)** in **b)** on **c)** of
10 He pleaded .... report it to the manager. **a)** me not to **b)** to me not to **c)** with me not to

3 Complete the following sentences by putting in one of the verbs from A and two of the prepositions from B to form a three-part phrasal verb. You can use any verb or preposition more than once. You may have to change the verb tense. (10 points)

A

look	come	face	get	break	run	cut	put

B

on	off	of	out	to	up	down	with	against	round

1 You shouldn't .... people just because they haven't had the same opportunities as you.
2 I know it's difficult but you just have to .... the fact that the business can't continue.
3 James .... a brilliant idea for a new slogan yesterday.
4 I've been meaning to sort out those files for weeks but I never seem to .... it.
5 We could .... our electricity bills by switching off our sign at night.
6 You'll never believe this! He's just phoned to say he can't make the meeting because he's .... petrol.
7 When we tried to introduce the clocking-in system we .... a lot of opposition.
8 It really is wrong that you ignore the rules about smoking and I'm not going to .... it any longer.

9 She really needs to .... her daily routine and do something completely different.

10 Andrew and I are good friends but I don't really .... his brother.

4 Choose the most suitable alternative to complete the sentences. (16 points)

1 You need to be there on the staff training day, just to make sure that everything runs ....
a) evenly b) flatly c) smoothly d) softly

2 The company won't be very pleased if we .... out of the deal at this stage.
a) pull b) fall c) run d) take

3 It took him a long time to come to .... with his redundancy.
a) rules b) words c) terms d) steps

4 I wish the managers wouldn't .... such unrealistic deadlines.
a) put b) lay c) place d) set

5 People say that it's easy to begin a job but it's harder to see it ....
a) off b) through c) away d) round

6 The meeting seemed to .... on for hours with no decisions being made.
a) draw b) pull c) drag d) haul

7 The company offers excellent working conditions and a very .... salary.
a) competitive b) contrasting c) comparable d) contradictory

8 I fully intended to sack him but in the end I couldn't .... through with it.
a) get b) put c) go d) run

9 She suffered a few disappointments at first but she took them all in her ....
a) pace b) march c) step d) stride

10 The course is intended for well qualified graduates who wish to improve their career ....
a) perspectives b) views c) prospects d) vistas

11 He started his first business .... as soon as he left school and now he's a millionaire.
a) affair b) essay c) venture d) trial

12 I was shocked to hear that Mrs Cane had decided to .... in her notice.
a) place b) hand c) put d) offer

13 It's a very competitive field, but if that's what you really want to do, then don't be .... off.
a) sent b) put c) set d) held

14 In these times of high unemployment everyone thought my giving up my job was .... madness.
a) sheer b) steep c) high d) deep

15 I don't know how I am going to cope .... working without my personal assistant.
a) in b) on c) thought d) with

16 I trusted you to do the job properly but you've .... me down.
a) let b) turned c) put d) kept

5 Complete the following sentences using the correct form of the word in bold. (6 points)

1 When I said I was starting up my own business, she looked at me in .... **BELIEF**

2 You can't do the job if you can't work in a team, ... of how many qualifications you have. **RESPECT**

3 Applicants need excellent organisational skills and the ability to work .... **DEPEND**

4 The lack of any reliable supply of electricity in that region came as a real .... to me. **REVEAL**

5 I think the female members of staff feel a bit .... when the men just talk about football. **ALIEN**

6 You need to be able to write with .... and precision. **CLEAR**

Unit 9 test

1 In the following text, put the verbs in brackets into the correct tense or form. Choose from: simple present, present continuous, simple past, past continuous, present perfect, past perfect, gerund and infinitive with *to*. (14 points)

Ever since I was a child, I (1) .... (always enjoy) mountain walking, but I think my best and most challenging trek was when I (2) .... (walk) the Inca Trail in Peru two years ago with my friend Chris. It was something we (3) .... (plan) to do for years and we weren't disappointed.

Our adventure started in Cusco, the Inca capital. Just after we arrived, I felt my heart (4) .... (beat) very fast, just as if I (5) .... (do) vigorous exercise. I was surprised as I (6) .... (never suffer) from altitude sickness before, but the feeling soon (7) .... (pass). In the morning, we drove from Cusco to Ollantaytambo to begin the trek. The journey usually (8) .... (take) three days. Unfortunately, Chris (9) .... (suffer) from a bad stomach when we (10) .... (set) out so we had to take things slowly that first day. As a result, we covered far fewer kilometres than we (11) .... (hope). The next day, however, he was much better and we managed (12) .... (walk) as far as the second pass. We arrived at Intipunku, the last place where you can camp, at the end of the third day. Then we spent the last day (13) .... (visit) the wonderful ruins at Machu Picchu. The mountain scenery of the whole walk is breathtaking and Machu Picchu is somewhere that everyone should visit in their lives if they can.

I'd love to go back to Peru and do another trek but I (14) .... (not plan) to for another two years or so. I need a break before doing something like that again.

2 Rewrite the following sentences beginning with *what* and emphasising the part of the sentence in italics. (8 points)

1 We need to know *whether the transport strike is going ahead.*
What ....

2 They discovered *a fallen tree across the road.*
What ....

3 He *advertised for a travelling companion on a website.*
What ....

4 He said that *the town was virtually dead during the winter.*
What ....

5 They *arrived at the waterfall* but then they were caught in a storm.
What ....

6 He enjoyed *visiting the theme park* most.
What ....

7 She *fell asleep on the train and missed her stop.*
What ....

8 *Someone stole the money from his backpack.*
What ....

3 Complete the following text. If there is a word in capitals, use it to form a word that fits the gap. If there is no word in capitals, put a preposition into the gap. (18 points)

Are you looking for a hotel with a really (1) .... **SPECTACLE** location? If so, why not book into an underwater hotel?

The world's first underwater hotel is Jules Lodge in Florida. To enter, guests scuba dive 21 feet beneath the sea and swim through an (2) .... **OPEN** at the bottom of the building. The hotel is not short (3) .... creature comforts and there are hot showers, a TV, phone and videos. But the best thing has to be the (4) .... **STUN** beautiful location. The hotel is surrounded (5) .... the Emerald Lagoon. It is (6) .... **POSSIBLE** not to be captivated (7) .... the sight of angelfish, parrotfish and barracudas peering in at the window and the whole hotel is covered (8) .... anemones and sponges. The (9) .... **AUTHENTIC** of the location is what sets it apart (10) .... similar venues such as amusement parks.

However, Jules Lodge is dwarfed by Hydropolis, a project in Dubai to build a luxury underwater hotel, complete (11) .... 200 suites, a ballroom and a cinema. The hotel is shaped like a bubble to provide maximum (12) .... **RESIST** to the pressure of the water and it has been compared (13) .... a giant turtle. The aims of the project are not restricted (14) .... making money through tourism. It is the (15) .... **INSPIRE** of Joachim Hauser, who is fascinated (16) .... marine life and deeply concerned (17) .... the devastating effect that pollution can have on it. The hope is that the hotel will make guests more aware (18) .... the vast underwater ecosystem which is currently under threat.

4 Choose the most suitable alternative to complete the sentences. (10 points)

1 We must leave the hotel by 7.30 as we have a very .... schedule.
a) narrow b) squeezed c) tight d) compressed

2 It was the first time that he had ever .... foot in America.
a) placed b) put c) laid d) set

3 He's not very organised and finds it hard to .... deadlines.
a) join b) gain c) meet d) win

4 Finding a place to change your currency can be a real ....
a) heartache b) headache c) backache d) tummyache

5 In the end, someone .... pity on him and lent him enough money to get home.
a) took b) put c) gave d) got

6 Learning to greet people in their own language can help to ....
a) call the tune b) plumb the depths c) break the ice d) kick the bucket

7 They .... off for London very early in the morning.
a) put b) gave c) left d) set

8 Don't pay the first price that you are asked as you are expected to ....
a) giggle b) haggle c) joggle d) wriggle

9 He looked surprised when I .... my ground and insisted that the figures were correct.
a) stood b) kept c) made d) fought

10 Don't worry about booking a hotel as I can .... for the night.
a) put you up b) see you off c) take you on d) set you down

Progress test 2 (Units 6–10)

1 Complete the second sentence so that it has a similar meaning to the first sentence, using the word given. You must use between three and six words, including the given word. (10 points)

1 You really should apply yourself to some serious studying now. **time**

It's high .... to some serious studying.

2 We were about to reach an agreement when you interrupted us. **point**

We .... an agreement when you interrupted us.

3 The release of his next novel is scheduled for the end of June. **due**

His next novel .... at the end of June.

4 He finds it hard to use the template on the computer. **difficulty**

He .... the template on the computer.

5 'I really think you should go to the police, Mary,' said Felix. **urged**

Felix .... to the police.

6 I'm sorry that I threw away the instruction leaflet now. **kept**

I wish .... the instruction leaflet now.

7 I am really surprised that you believed his story. **surprises**

What .... you believed his story.

8 She persuaded Nicholas not to take early retirement. **talked**

She .... early retirement.

9 We won't achieve anything if we stand around like this. **point**

There is .... like this.

10 I find his attitude towards modern education the most irritating part of his work. **irritates**

The thing .... his work is his attitude to modern education.

2 In some of the following sentences, there is an incorrect verb form. Either correct the mistake, or if the sentence is already correct, put a tick at the end. (10 points)

1 The show was so funny that I couldn't stop to laugh. ....

2 I am looking forward to meet you on Tuesday. ....

3 Even as a child, he longed to visit Italy. ....

4 I can't imagine wanting to sell that painting. ....

5 They don't let you to touch any of the exhibits. ....

6 I am used to live in a hot country. ....

7 The news made me to feel quite depressed. ....

8 You'd better not criticising something you don't understand. ....

9 The painting sold well and she went on to hold her first exhibition. ....

10 I really miss to be able to meet up on Sunday afternoons. ....

3 In the following sentences some of the words in bold are incorrect as they cannot be used in the plural form. Rewrite these sentences, correcting the plural form and making any other necessary changes. If the sentence is already correct, put a tick underneath. (10 points)

1 Relatively few **researches** have been done into the ecosystem of the canopy.

....

2 Could you give me some **advices** on how to look after this plant?

....

3 There are not many **evidences** to suggest that wolves pose a serious threat to cattle.

....

4 The committee have made a number of **recommendations** for protecting the remaining Bengal tigers.

....

5 The storm did a great many **damages** to the new stadium.

....

6 The police found no **proofs** that the animals on the farm were being mistreated.

....

7 The **authorities** have imposed a complete ban on the ivory trade.

....

8 I can't bear people who drop their **litters** on the street.

....

9 He has had a number of frightening **experiences** while on safari.

....

10 We may spot the famous condor on our trek but don't raise your **hopes** too high.

....

4 Some of the following sentences are incorrect because there is a missing *it*. Insert *it* in the correct place if necessary. If the sentence is already correct, put a tick at the end of the line. (10 points)

1 I have already made clear that I am not interested in this project. ....
2 It is hard to justify the killing of animals for their fur. ....
3 Global warming poses a real threat to our planet and we must take seriously. ....
4 I find appalling that geese are treated so cruelly just to produce liver pate. ....
5 It is made quite clear in the rules that tenants are not allowed to keep pets. ....
6 We owe to future generations to look after our natural environment. ....
7 The survey found that most people would buy organic fruit and vegetables if they were cheaper. ....
8 Bull fighting may be a traditional sport but that does not make right. ....
9 I doubt whether they will allow a supermarket to be opened in such a rural area. ....
10 I take that you have heard the latest news about the campaign. ....

5 In the following text, choose the correct link word to fill each space. (14 points)

SAVE OUR BANANAS!

Bananas are one of the world's most popular fruit. (1) .... they are seen mainly as a dessert in most developed countries, they have many qualities which make them an ideal food. (2) ...., they are a rich source of many nutrients such as carbohydrates, vitamins and potassium. (3) ...., they are easy to eat because they contain no seeds.

It was not always like this, (4) .... Before they became a cultivated crop, wild bananas were inedible fruit, packed full of stony seeds. Edible bananas first arose as sterile mutants, which early farmers propagated by taking cuttings. Unfortunately, (5) .... their sterility, modern bananas are very vulnerable to disease. (6) ...., any disease or pestilence would be able to rip through an entire plantation with nothing to stop it (7) .... the bananas all share the same genetic make-up.

There are several such diseases which pose a threat to banana crops, (8) .... the most worrying is probably so-called Panama disease. In the 1950s, this fungus virtually wiped out the banana plantations in Central America and the Caribbean. The local economy only survived (9) .... the fortunate discovery of a type of banana plant called Cavendish, which was resistant to the disease. These Cavendish bananas have become the standard commercial variety available today.

(10) .... the problems are not over. In 1992, Panama disease mutated into a form which is able to attack the Cavendish variety. It has already had devastating effects on the plantations in Malaysia and is a serious threat to banana production in Latin America. (11) ...., this new strain of disease is also able to attack plantain, an important staple food in that continent.

Developing a new disease-resistant variety of banana is (12) .... a matter of some urgency. Significant progress has been made. A new strain, known as the FHIA banana, is currently being grown in parts of Africa and the Caribbean. It is resistant to all major diseases and is (13) .... highly productive. (14)....... it may be difficult to persuade the global consumer to switch to this new type of banana. According to some people, it has a distinct taste of apple!

	A	B	C
1	A But	B Although	C However
2	A However	B In addition	C To begin with
3	A Firstly	B Secondly	C Nonetheless
4	A however	B furthermore	C although
5	A since	B because of	C because
6	A However	B Moreover	C Nevertheless
7	A since	B but	C so
8	A but	B so	C as
9	A due to	B in spite of	C as well as
10	A So	B But	C Even
11	A Consequently	B In the first place	C Furthermore
12	A therefore	B so	C however
13	A also	B firstly	C as well as
14	A So	B Therefore	C Yet

6 Complete the following sentences with the correct prepositions. (10 points)

1 He insisted .... keeping expenses as low as possible.
2 We set .... on our journey at 8 o'clock.
3 His opinion is very similar ...... mine.
4 Congratulations ...... your promotion!
5 Entrance is restricted .... members only.
6 Let's hope the election result will bring .... a few changes.
7 There are a few problems but try to look .... the bright side.
8 He apologised ...... speaking so thoughtlessly.
9 I was never very good ...... Maths at school.
10 The policy has resulted ...... a wider gap between the rich and the poor.

7 Complete the following sentences with one of the verbs from the box, plus either the particle *up* or *down*. (8 points)

mount	water	narrow	stir	cut	brush	die	track

1 If more people used public transport this would help to .... on greenhouse gas emissions.
2 In their attempt to catch the thieves, the police have managed to .... their investigation to just three suspects.
3 We must insist that the government keep to their promise to introduce a carbon tax and not allow them to .... the policy.
4 The campaign against the new runway has managed to .... feelings among most of the local people.
5 Thousands of people demonstrated against the new laws and I do not expect their anger to .... for a long time.
6 The police have managed to .... a group of four local men who are accused of providing false passports.
7 We could hire a solicitor to help us with our case, but I don't want the legal fees to ....
8 You'll need to .... your Spanish before you go to South America.

8 Each of the following sentences contains one word which is incorrectly used. Cross it out and write a correct word. (10 points)

1 There was no inspector on the train so nobody controlled the tickets at all. ....
2 They were not expecting to loose the match. ....
3 The vase turned out to be priceless so I gave it to the charity shop. ....
4 As usual, it is the most vulnerable people who will be effected by the cutbacks. ....
5 It's a good idea in principal but it won't work in practice. ....
6 The dove appears in the painting as a logo of peace. ....
7 Is there any chance that they will rise our salary? ....
8 All of his clothes were laying in a heap on the floor. ....
9 Next month, the theatre company will display a new production of *Macbeth*. ....
10 It is surprising how much dirt can hoard inside the computer mouse. ....

9 Choose the most suitable alternative to complete the sentences. (12 points)

1 Please .... an eye on the dog and see he doesn't go on to the road.
a) hold b) keep c) place d) put
2 We have to face the truth about the ecological crisis and stop burying our heads in the ....
a) earth b) soil c) sand d) dirt
3 In many countries you cannot buy antibiotics without a .... from the doctor.
a) receipt b) chit c) invoice d) prescription
4 The white rhino is still on the .... of extinction despite the efforts to save it.
a) edge b) brim c) border d) verge
5 For many years, Chico Mendes was at the .... of the movement to save the Amazon forest.
a) forehead b) forefront c) foreground d) forehand
6 He has a number of health problems but he tends to make ...... of them.
a) gentle b) light c) soft d) shallow
7 Adult sparrows require a good supply of live food on which to .... their young.
a) rear b) grow c) breed d) lift
8 I think things may well improve under this government but only time will ....
a) say b) show c) tell d) display
9 Large areas of the forest have been cut down and the land has been ...... over to cattle farming.
a) given b) put c) taken d) made
10 Unfortunately, I think you've ....up an infection from looking after the birds.
a) taken b) picked c) lifted d) caught
11 The hawk flew over to the .... of trees at the end of the field.
a) bunch b) clump c) range d) bundle
12 The farm has been restocked with goats, chickens and a small .... of cattle.
a) flock b) pack c) swarm d) herd

10 What type of text do the following sentences refer to? Choose from A, B or C. For some sentences you will need to write two letters. (6 points)

A *report* B *proposal* C *review*

1 The introduction is written in a way which is intended to catch the reader's attention.
2 It is always written in formal language.
3 It puts forward a new idea with specific plans for the future.
4 The introduction states its aim or purpose.
5 The writer's opinion is only given at the end.
6 It uses a range of descriptive and often idiomatic vocabulary.

Unit 11 test

1 Complete the second sentence so that it has a similar meaning to the first sentence, using the word given. Do not change the word given. You must use between three and six words, including the given word. (15 points)

1 He has a bad leg so I know he has not gone far. **have**
He .... with his bad leg.

2 Going out without your coat and scarf was foolish. **should**
You .... your coat and scarf before going out.

3 It's possible that he was driving the car when you phoned him. **have**
He .... the car when you phoned him.

4 It was wrong of you to lie to me. **to**
You .... the truth.

5 Maria says that it is impossible that he knew about the meeting. **way**
According to Maria, there is .... about the meeting.

6 As soon as they had closed the door, the alarm went off. **sooner**
No .... the door than the alarm went off.

7 He writes very original poetry and is a mathematical genius as well. **only**
Not .... very original poetry, he is a mathematical genius as well.

8 The president never admitted that his strategy had failed. **time**
At .... that his strategy had failed.

9 You should never reveal your password to anyone else. **circumstances**
Under .... your password to anyone else.

10 As soon as his lecture came to an end, the students broke into applause. **finished**
Hardly .... when the students broke into applause.

11 People only began to realise how intelligent she was after she had left school. **start**
Only after she had left school .... how intelligent she was.

12 Our memory of the words that someone has used is not usually exact. **exactly**
Seldom .... what words someone has used.

13 It only dawned on them who he was when he started speaking to them. **realise**
Not until he started speaking to them .... who he was.

14 They don't often spend much money on cultural activities. **rarely**
Only .... on cultural activities.

15 Parents should not let children watch TV before the age of three. **allowed**
Only after the age of three .... watch TV.

2 Choose the most suitable alternative to complete the sentences. (15 points)

1 If you don't mind, I'd like to .... your brains about writing a business plan.
a) dig b) pick c) harvest d) gather

2 I'd advise you to take any criticism in your .... and just think of it as a learning opportunity.
a) pace b) steps c) stride d) walk

3 I think I'm quite easy to get on with but I do tend to speak my ....
a) thought b) mind c) brain d) idea

4 One of the keys to happiness is to appreciate the little things in life and not take them for ....
a) given b) assumed c) granted d) awarded

5 I've been .... my brains all weekend to think of a possible solution.
a) racking b) stretching c) squeezing d) wringing

6 It's time you stopped daydreaming and .... up to reality. a) fronted b) faced c) headed d) turned

7 He's her only grandchild and she thinks the .... of him. a) earth b) globe c) planet d) world

8 Fancy walking all this way in the rain without even an umbrella! Are you .... your mind?
a) out of b) away from c) apart from d) off of

9 Ever since we advertised our course on improving your memory, we have been .... with enquiries.
a) invaded b) assaulted c) bombarded d) peppered

10 I do know her husband's name but it's just .... my mind. a) escaped b) slipped c) evaded d) flown

11 The job requires excellent communication skills and an ability to think on your ....
a) toes b) feet c) hands d) legs

12 I didn't agree with all of his argument but it was certainly thought ....
a) arousing b) inciting c) provoking d) encouraging

13 Talking to someone who has had a similar experience may help you come to .... with your loss.
a) conditions b) terms c) contract d) agreement

14 There are many activities for toddlers, but no timetable as they have such a short attention ....
a) stride b) span c) length d) space

15 I remember having to learn poems by .... at school. a) heart b) mind c) memory d) head

3 Complete the following sentences with the correct prepositions. (10 points)

1 I can't concentrate .... my work at all while you're making that noise.
2 I think you need an evening out to take your mind .... things.
3 I'm not sure whether to accept the offer, but I'll think it .... at the weekend.
4 I know your situation isn't perfect, but try to look .... the bright side.
5 He seems to be completely devoid .... any sense of guilt at what he's done.
6 He was too preoccupied .... his own thoughts to notice who was standing next to him.
7 We plan to take .... three new members of staff next month.
8 You didn't make a very good impression .... my parents, I'm afraid.
9 It was the first time he had forgotten the homework, so the teacher let him ....
10 I met the new boss last week but I can't say I really took .... her.

4 Complete the following sentences using the correct form of the word in bold. (5 points)

1 Whatever happens, we'll just have to accept it .... **PHILOSOPHY**
2 If you are as happy as you claim, then you have reached a truly .... state. **ENVY**
3 Everyone is aware of how good you are at your job so you shouldn't feel so .... **APPRECIATE**
4 He didn't know how to work the camera and just looked at it rather .... **HELP**
5 Memories of .... experiences from childhood can stay with you for ever. **TRAUMA**

5 Complete each sentence with the correct word from the box. (5 points)

recall	memory	memorise	mind	remind

1 Can you .... me to buy a new toner cartridge this afternoon?
2 I can't .... ever meeting him before.
3 She is prone to sudden lapses of ....
4 You shouldn't write down your PIN number; you just have to .... it.
5 He didn't say anything but I could tell there was something on his ....

Unit 12 test

1 Complete the second sentence so that it has a similar meaning to the first sentence, using the word given. Do not change the word given. You must use between three and six words, including the given word. (10 points)

1 He liked people to think of him as a strong leader. **thought**
He liked .... as a strong leader.

2 They made previous generations of village children help with the harvest. **were**
Previous generations of village children .... with the harvest.

3 It is estimated that over 5000 people were present at the coronation. **attended**
Over 5000 people are .... the coronation.

4 No one can decide anything until we hear more news. **be**
No decision .... until we hear more news.

5 People do not feel that the committee has done a satisfactory job. **felt**
The committee .... a satisfactory job.

6 They say that Socrates died from drinking hemlock. **said**
Socrates .... from drinking hemlock.

7 We have to plan for a possible influx of refugees. **made**
Plans .... for a possible influx of refugees.

8 People think that someone started the fire on purpose. **have**
The fire .... on purpose.

9 The original plan was to convert the castle into a hotel. **have**
In the original plan, the castle .... into a hotel.

10 They heard him say that the king was seriously ill. **was**
He .... that the king was seriously ill.

2 Choose the most suitable verb forms to complete the following text. (10 points)

The Millennium Dome is a famous dome-shaped building in South East London. It (1) *was intended to be / would be / would have been* a monument to celebrate the coming of the new millennium and, in the original proposal, it (2) *was to become / was becoming / would become* one of the most famous and frequently visited sites in the city. Many politicians expressed their optimism about it, including Tony Blair who claimed that it (3) *will become / would become / would have become* 'a beacon to the world'.

Unfortunately, the dome failed to live up to such high hopes. It was predicted that it (4) *will attract / would attract / would have attracted* about 12 million visitors but the real number was only half of this. In fact, if that many visitors really had come, then there (5) *will be / would be / would have been* severe problems with overcrowding and long queues. During 2000, the organisers continually requested more money for the project and although changes were made at management level which (6) *were improve / were going to improve / were aimed at improving* the financial management, they actually made very little difference.

The dome was closed at the end of 2000. A bid to purchase the site was put in by a business group which (7) *was planning to turn / would turn / were turning* it into a business park but just as the deal (8) *would be struck / was about to be struck / would have been struck*, the government decided that they were unhappy with it. Finally in May 2002, it was announced that the building (9) *was to become / was becoming / would have become* an entertainment centre. There are new plans to hold sporting events there and it is now confirmed that it (10) *will be used / would be used / would have been used* to host the gymnastics and the basketball matches for the 2012 Olympics.

3 Complete the following sentences using the correct form of the word in bold. (8 points)

1 This has definitely been one of the most .... experiences of my life. **MEMORY**
2 He plans to spend the summer working on an .... site in Greece. **ARCHEOLOGY**
3 The museum offers many interactive and computer-based activities as well as .... information about the town. **HISTORY**
4 There was widespread .... that the king was planning to abdicate. **SPECULATE**

5 The book is a collection of .... of his childhood in a Cornish village. **REMINISCE**

6 Convicted criminals have always been .... from voting in elections. **ALLOW**

7 Architects are designing a new .... building as part of the development of the town centre. **FUTURE**

8 Although his ideas were dismissed during his own time, he is now often regarded as a .... **VISION**

4 Choose the most suitable alternative to complete the sentences. (12 points)

1 We cannot .... out the possibility that the whole plot was engineered by the government of the time.
a) strike b) rub c) rule d) cross

2 With the benefit of ...., the king's decision to send such a small army seems very optimistic.
a) foresight b) hindsight c) short-sight d) long-sight

3 His latest novel is .... in France at the time of the revolution. a) placed b) set c) put d) laid

4 I am sure the system will work in the end but there are still a great many glitches to .... out.
a) wash b) pick c) iron d) shake

5 The company was planning to open a new office but it fell .... due to lack of money. a) out b) over c) through d) under

6 Travelling from one town to another was a dangerous activity in those days, mainly because of the bands of robbers .... along the road. a) lurking b) lounging c) lurching d) larking

7 The leader .... a different role to each member of the team. a) attributed b) distributed c) assigned d) ascribed

8 His historical novels are entertaining but he has a very poor .... of the politics of the time.
a) grip b) hold c) clutch d) grasp

9 The prime minister was furious when details of the inquiry were .... to the press. a) dripped b) leaked c) trickled d) gushed

10 She's always .... back to the time when she lived in that little village. a) hailing b) harking c) harping d) hurling

11 The fact that there is no contemporary record of his death gives .... to the theory that he actually fled the country.
a) height b) depth c) length d) weight

12 He never made a career as an artist but continued to .... in different forms of painting all his life.
a) dabble b) dribble c) dawdle d) dangle

5 For the following questions think of **one** word only which can be used appropriately in all three sentences. (5 points)

1 After the owner's death, the castle became the property of the state and in due.........., it was opened to the public.
The meeting of the two world leaders changed the .... of history.
He decided to take a refresher .... to improve his driving.

2 She left the company of her own free .... and no one put any pressure on her.
They confessed to the crime but afterwards they claimed that they had been forced to against their ....
You have to stick to your diet to lose weight, which can take a lot of .... power.

3 There is a theory that political events in one place will often .... off similar events in others.
The .... for the riots was the government's proposal to reduce wages.
He took out the pistol, took aim and squeezed the ....

4 The book has a very complicated .... which is difficult to follow on the first reading.
In 1605, a group of conspirators hatched a desperate .... to murder the entire government by blowing them up with gunpowder.
They used a line graph to .... the patients' reactions to the new drug.

5 We cannot know what the future may .... so it's a waste of time worrying about it.
They decided to .... a special meeting to discuss the new project.
They managed to escape from the building just before the fire really took ....

6 In the following sentences, **two** of the expressions are possible and **one** is not. Cross out the incorrect alternative. (5 points)

1 I'd love to stay and chat longer but I'm afraid I'm a bit .... a) pressed for time b) behind the times c) behind time

2 I can't move into my new flat until next week so I'm staying with a friend ....
a) in time b) for the time being c) for the moment

3 I had an hour to wait for the bus so went to look around the bookshop to .... a) make time b) pass the time c) kill time

4 I've warned you about that .... but you never take any notice. a) time after time b) in the long run c) time and time again

5 The fire was spreading towards the children's bedroom but the fire brigade arrived ....
a) just in time b) in the nick of time c) right on time

Unit 13 test

1 In the following sentences, a word is missing. Insert a suitable word in the correct place. There is one sentence which has no missing word. (8 points)

1 Spent so much time preparing her presentation, she was upset when the seminar was cancelled.
2 Don't write too much because answers exceed the given word limit will not be marked.
3 Rather forgetful, she left the house without her purse or her glasses.
4 Dyirbal is one of a large number of Aboriginal languages spoken in Australia.
5 Children take regular exercise do better academically too.
6 Manx has now disappeared as a first language, its last native speaker having in the 1960s.
7 Having brought up bilingually, he was the obvious candidate for the interpreting job.
8 James Murray was knighted in 1908, worked on the *Oxford English Dictionary* all his life.

2 Complete the second sentence so that it has a similar meaning to the first sentence, using the word given. Do not change the word given. You must use between three and six words, including the given word. (8 points)

1 There have been many debates about the origin of the language recently. **much**
The origins of the language .... recently.
2 People think that the letter was written in a secret code. **been**
The letter .... in a secret code.
3 The police brought charges against him for wasting their time. **charged**
He .... police time.
4 Latin and Greek no longer form part of the curriculum. **dropped**
Latin and Greek .... the curriculum.
5 She doesn't like it when people tell her what to do. **told**
She doesn't take kindly .... what to do.
6 English is now widely used instead of Yoruba as the language of West African universities. **replaced**
Yoruba .... English as the language of West African universities.
7 They are doing all they can to ensure that the language does not die out. **being**
All possible measures .... the language from dying out.
8 There will be a clear reduction in productivity if they adopt this policy. **should**
There will be a clear reduction in productivity .... adopted.

3 Choose the most suitable adverbial in the following exchanges. (6 points)

1 'She had her first Chinese lesson yesterday but I don't think she liked it much.'
'Why?'
'Well, ... all they did was copy notes off the board.' **a)** clearly **b)** apparently **c)** personally
2 I don't think changing the timetable is a very good idea, ... Everyone seems perfectly happy with it as it is.
a) actually **b)** obviously **c)** apparently
3 'What was Skinner's view of language acquisition?'
'Well, ..., he believed that children learn language by copying what they hear.'
a) frankly **b)** personally **c)** basically
4 'Did you enjoy the film?'
'I think Andy did but ...., I found it rather dull.' **a)** apparently **b)** clearly **c)** personally

5 'You could have told me you'd be late. Didn't you have your mobile phone with you?'
'Well, ..., I didn't or I would have called you.' a) obviously b) apparently c) frankly

6 'Putting it quite ...., you have no chance of passing your exam.' a) basically b) obviously c) frankly

4 Complete the following sentences using the correct form of the word in bold. (10 points)

1 The building is checked .... to make sure that there is no structural weakness. **PERIOD**
2 The new edition of our dictionary is .... all its competitors. **SELL**
3 Many people are only now starting to realise, somewhat ...., how serious the threat of global warming is. **LATE**
4 Many people consider him one of the finest writers of the time, but personally I think he's rather .... **RATE**
5 It has been suggested that offering schoolchildren a more flexible timetable might reduce the .... rates. **TRUANT**
6 The grammar of most Slavic languages is tough although the spelling system is .... easy. **COMPARE**
7 We really appreciate all your hard work and your .... **PROFESSION**
8 The .... of self-study resources is very important to ensure that students can work effectively outside class. **AVAILABLE**
9 That child has been expelled from school for .... behaviour. **DISRUPT**
10 The proposal may tackle the problem superficially but it does not really address the .... issues. **LIE**

5 Choose the most suitable alternative to complete the sentences. (18 points)

1 He's studied oriental languages and .... to a career in the diplomatic service.
a) aspires b) pursues c) projects d) intends
2 She was given permission to hand in her thesis late on medical ....
a) principles b) motives c) grounds d) excuses
3 The new language-learning CDs are intended to .... into the demand for self-study courses.
a) tap b) dig c) cut d) drill
4 I'm just starting to get to .... with the language although it is very difficult.
a) grasp b) grips c) clutches d) clasp
5 He listened to the lecture with .... concentration. a) heavy b) strong c) high d) intense
6 I was so tired last night that I slept like a .... a) stick b) plank c) log d) branch
7 I've never taken any Spanish lessons but I .... up quite a lot when I stayed in Spain.
a) took b) picked c) pulled d) gathered
8 She may speak her .... more bluntly than some people would like but at least she's honest.
a) brain b) thought c) mind d) idea
9 The lecture was so complicated I could not make .... of it.
a) head nor tail b) hand nor foot c) back nor front d) arm nor leg
10 It's an unconventional idea but a really good example of thinking out of the .... a) pack b) case c) box d) bag
11 I thought she would be nervous about her oral exam but she went in cool as a ....
a) lettuce b) cucumber c) tomato d) radish
12 I wish you'd stop rambling and get to the .... a) tip b) point c) edge d) nib
13 He wanted to go to teach in Japan but his application was .... down. a) sent b) put c) set d) turned
14 I tried to explain it wasn't my responsibility but she got the wrong end of the .... and thought I was saying it was hers.
a) pole b) branch c) post d) stick
15 I never feel very comfortable at parties where you have to make .... talk with people you don't know.
a) little b) light c) small d) short
16 The government will start to .... the new policy on education next year.
a) implement b) involve c) stage d) configure
17 It's the beginning of the weekend and we're in the bar so let's not talk ....
a) office b) shop c) stall d) market
18 He talks on and on and never lets you get a word in ....
a) sideways b) backwards c) edgeways d) upwards

Progress test 3 (Units 11–14)

1 Complete the second sentence so that it has a similar meaning to the first sentence, using the word given. Do not change the word given. You must use between three and six words, including the given word. (16 points)

1 It's possible that he was waiting for us in the wrong place. **could**

He .... for us in the wrong place.

2 It is never a good idea to wander away from the path. **should**

Under .... wander away from the path.

3 I am sure that it wasn't Roger that you saw at the party. **have**

It .... Roger that you saw at the party.

4 I had only just picked up the receiver when I heard the ambulance arrive. **lifted**

Hardly .... I heard the ambulance arrive.

5 People believe that Alan's wife destroyed his diaries after he died. **have**

Alan's diaries .... by his wife after he died.

6 The moment that they sat down, the chairman walked in. **sooner**

No .... the chairman walked in.

7 The idea first came to her when she was working as a waitress. **struck**

She .... the idea when she was working as a waitress.

8 Both his parents were actors so he is very familiar with the world of the theatre. **having**

He is very familiar with the world of the theatre, ... actors.

9 The original idea was to complete the restructuring by the beginning of March. **have**

According to the original plan, the restructuring .... by the beginning ot March.

10 It is quite possible that the emergency call was a hoax. **well**

The emergency call .... a hoax.

11 No one can help the injured until the rescue team arrive. **can**

There is nothing .... the injured until the rescue team arrive.

12 I was tricked into giving out my bank details before so it won't happen again. **having**

I won't give out my bank details again, ... it before.

13 Nobody took any notice of safety standards before the laws were passed. **attention**

No .... safety standards before the law was passed.

14 You shouldn't assume that the committee will support your proposal. **granted**

You shouldn't .... the committee will support your proposal.

15 They were just going to sign the contract when David spotted a mistake. **about**

The contract .... when David spotted a mistake.

16 You should make sure that you benefit from the company's in-house training. **take**

You should .... the company's in-house training.

2 Each of the following sentences contains one word which is incorrectly used. Cross it out and write a correct word. (8 points)

1 Martin left home an hour ago so he must has arrived home by now. ....

2 Not only you borrowed my car without permission but you scratched it too. ....

3 He mustn't have been wearing his life jacket or he would have survived. ....

4 Only after he had left the room I realised what he had said. ....

5 They would build a new railway but the plan was halted because of protests from environmental groups. ....

6 The effects of the new policy will monitor closely over the first 12 months. ....

7 We cannot allow mistakes like this to make again. ....

8 The bathers watched the waves were crashing against the rocks. ....

3 Complete the following sentences using the correct form of the word in bold. (10 points)

1 Some people believe that Echinacea is .... if you are going down with a cold. **BENEFIT**
2 Our aims are simple: to reduce costs and .... profits. **MAXIMUM**
3 Despite free education, there has been little change in social .... **MOBILE**
4 It is an interesting idea but I do not believe it is .... viable. **COMMERCE**
5 They were not prepared for the flood and could only watch .... as the water came into the room. **HELP**
6 Local residents raised a number of .... to the proposed building project. **OBJECT**
7 The new brochures can be produced at a .... low cost. **COMPARE**
8 Although the doctors had been optimistic, the patient's condition actually .... during the night. **WORSE**
9 With ...., it is easy to see that more preparations should have been made for an emergency. **SIGHT**
10 I am .... to go to the police unless we have any real evidence. **WILL**

4 Find and correct one spelling mistake in each of the following sentences. Two of the sentences have **no** spelling mistakes, in which case put a tick at the end. (13 points)

1 I have still not recieved a reply to my enquiry.
2 I am dissappointed by the lack of interest in the topic.
3 She never accepts any responsability for these bad decisions.
4 The hotel can acommodate over 90 visitors at one time.
5 There has been a definite improvement in the standard of your writing.
6 It is usefull to keep the receipts in two separate piles.
7 Some people do not believe that accurate spelling is necessary.
8 They had a furious argument in public, which was embarassing for everyone else.
9 I would reccommend the fruit salad for dessert.
10 Unfortunatly, vandalism is a frequent occurrence in this street.
11 I don't normaly like detective stories but that one is a masterpiece.
12 The book is an enjoyable read althought the plot is rather predictable.
13 Wheather Harry Potter is truly a children's classic is still debatable.

5 Correct the punctuation in the following sentences. In each sentence you must either **add** or **remove** a punctuation mark. (13 points)

1 All of the book's that he wrote turned out to be best-sellers.
2 People say, that you should never judge a book by its cover.
3 It is probably the best of his novels however, it is not so well known as some of the others.
4 Helen White, whose first novel was published in 2002 has just won an award.
5 It was a long, boring, book, written in a turgid style.
6 Despite popular tradition Sherlock Holmes never says, 'Elementary, my dear Watson!'
7 Both of the students essays were outstanding pieces of work.
8 It is a carefully written account, but it's general thesis is out of date.
9 What really impressed me about the book, was its attention to detail.
10 Shakespeare's plays can be divided into three groups comedies, tragedies and histories.
11 There are different theories about who wrote the sequel to the poem?
12 He keeps saying that the pen is his, but I'm sure that it's our's.
13 He wrote his first novel, *Hanging Man*, at the age of 17 which was a remarkable achievement.

6 Choose the most suitable alternative to complete the sentences. (20 points)

1 I read the .... on the back cover of the book, and I decided not to buy it.
a) glossary b) blurb c) index d) footnote
2 I've got so much paperwork to .... through that it will take me all weekend.
a) swim b) wade c) float d) paddle
3 The book has a complicated plot with a surprising .... at the end.
a) turn b) bend c) twist d) spin
4 I just .... a glimpse of the castle as the coach turned the corner.
a) caught b) took c) received d) stole
5 He .... around the bookshop for over an hour before finally choosing a book.
a) skimmed b) perused c) flipped d) browsed
6 After the success of his first novel, the second did not really .... up to people's expectations.
a) put b) stand c) live d) make
7 The collection contains some very interesting articles which really .... your attention.
a) snatch b) grab c) pluck d) clutch

8 If you don't have time to read a long novel, just .... into his collection of short stories.

a) dip b) dive c) drop d) dunk

9 The lyric poetry of Ancient Greece only survives in a few ....

a) scraps b) tatters c) fragments d) shards

10 The reader is gradually .... into the story by the clever distribution of clues.

a) drawn b) drafted c) pulled d) hauled

11 The sound of the fire alarm made me jump out of my ....

a) shape b) body c) frame d) skin

12 If you are looking for an expert in this field, then the name Professor Franklin immediately .... to mind.

a) jumps b) hops c) springs d) bounds

13 She spent the whole weekend agonising .... whether to take the job or not.

a) over b) around c) through d) across

14 You can't borrow reference books in the normal .... of things, but we might make an exception in this case.

a) run b) walk c) drive d) flight

15 The book gives a .... outline of the main changes that took place during the period.

a) wide b) large c) thick d) broad

16 The investigation seems to be going nowhere but then the detective .... upon an important clue at the last minute.

a) trips b) stumbles c) slips d) sprawls

17 He stood outside the boss's office, ... nervously with a button on his shirt.

a) meddling b) fiddling c) tampering d) fidgeting

18 The special effects in that film are out of this ....

a) world b) earth c) globe d) planet

19 They say that some .... in life are so incredible that you could never put them in a book.

a) concurrencies b) collaborations c) conformities d) coincidences

20 It's not a book that you need to read from ...., so just pick out the relevant parts.

a) page to page b) front to back c) cover to cover d) top to bottom

7 Decide which of the following statements about the CAE exam are true and which are false. (20 points)

1 The Reading paper lasts exactly one hour.
2 In Part 2 of the Reading paper (gapped text), the text is always a narrative.
3 In Part 3 of the Reading paper (multiple choice), the questions follow the order of the information in the text.
4 In the Reading and Listening papers, marks are deducted for wrong answers.
5 Part 1 of the Writing paper always involves writing a letter.
6 In the Writing paper, your answer to both parts should be 220–260 words long.
7 Question 5 of the Writing paper always involves writing about one of the set books.
8 Part 1 of the Use of English paper (multiple-choice cloze) only tests vocabulary, not grammar.
9 In Part 2 of the Use of English paper (open cloze), you can put two possible answers in a gap.
10 In Part 3 of the Use of English paper (word formation), some of the words may not need to be changed.
11 In Part 5 of the Use of English paper (key word transformations), you must not change the key word in any way.
12 In the Listening paper, you will hear all of the listening passages twice.
13 At the end of the Listening paper, you will have five minutes to copy your answers onto the answer sheet.
14 In Part 2 of the Listening paper, if an answer is incorrectly spelt, you can still get the mark.
15 In the Reading and Listening papers, you should write your answers on the answer sheet in pencil.
16 There are four parts to the Speaking exam.
17 In the Speaking exam, both examiners will talk to you, but at different stages of the test.
18 In Part 2 of the Speaking exam, you should only describe exactly what you see in the picture.
19 You will be marked down in the Speaking exam if you do not allow your partner to talk.
20 At the end of the Speaking test, the examiner will tell you if you have passed or failed.

Test keys

Unit 1 test

1 1 haven't seen 2 have you been doing 3 arrived 4 had already started 5 weren't singing 6 had been miming 7 will have performed 8 will have played 9 has had 10 will survive

2 1 complimentary, performance 2 inspiration 3 professional, dedication 4 popularity 5 thunderous, appearance 6 creative, successful 7 arguably 8 encouragement

3 1 show's greatest weakness/greatest weakness of the show 2 a wonderfully accomplished 3 general reluctance among 4 reacted immediately by sacking 5 more innovations than

4 1 top 2 cover 3 pitch 4 note 5 peak

5 1 c) 2 b) 3 d) 4 a) 5 b) 6 c) 7 a) 8 d) 9 d) 10 b)

6 1 complain *about* 2 ✓ 3 concentrate *on* 4 crazy *about* 5 ✓ 6 owing *to* 7 ✓ 8 consists *of*

Unit 2 test

1 1 (-) 2 an 3 the 4 (-) 5 a 6 the 7 a 8 (-) 9 (-) 10 the 11 the 12 a 13 (-) 14 the 15 (-)

2 1 who 2 whose 3 that 4 which 5 where

3 1 that the 2 generations, was 3 its 4 money that 5 to last.

4 1 safe 2 far 3 level 4 high 5 account

5 1 consumption 2 disloyalty 3 unwanted 4 attachment 5 memorable 6 discernment 7 insecurity 8 economise 9 unattractive 10 potentially

6 1 c) 2 b) 3 a) 4 d) 5 c) 6 a) 7 c) 8 a) 9 b) 10 c)

Unit 3 test

1 1 to limit 2 a) 3 c) 4 demanding 5 a) 6 b) 7 to put 8 to draw 9 spending 10 a) 11 to move 12 to work 13 to forgive 14 a) 15 a) 16 b) 17 c) 18 to look

2 1 had to take 2 you ought to contact 3 do not/don't have to 4 might like to 5 should have known better 6 went on/carried on taking 7 allow us/you/people to recognise 8 should avoid arguing 9 am not prepared to listen 10 force him to go

3 1 conscientious 2 taciturn 3 quick-tempered 4 arrogant 5 self-conscious 6 inquisitive 7 gullible 8 trustworthy 9 vindictive 10 high-spirited

4 1 c) 2 b) 3 d) 4 b) 5 a) 6 b) 7 b) 8 b) 9 d) 10 c) 11 a) 12 b)

Unit 4 test

1
1 had not been given
2 were to send clear/clearer
3 should you require
4 checks had been carried out
5 I to meet the inventor
6 you were looked after/you were being looked after
7 if you happen to know
8 I realised how unreliable it was
9 parents were given the possibility
10 we were to invest more

2
1 if we ~~will~~ start cultivating …
2 would not *have* put
3 Unless we are ~~not~~ responsive
4 might never have *been* invented.
5 If you happen *to* see James
6 Had the CCTV *been* working
7 provided ~~with~~ it is serviced
8 If we were *to* concentrate
9 If I had ~~have~~ known
10 If it had not been *for* the discovery

3 1 ~~control of something~~ 2 ~~a question~~ 3 ~~issue~~ 4 ~~charge of something~~ 5 ~~quiet~~ 6 ~~a favour~~ 7 ~~a breath~~ 8 ~~a picture~~ 9 ~~your tongue~~ 10 ~~a difference~~

4 1 matter 2 twist 3 rule 4 pack 5 shred

5 1 sense 2 root 3 case 4 blue 5 science

6 1 c) 2 a) 3 c) 4 b) 5 b) 6 d) 7 b) 8 c) 9 b) 10 d)

Progress test 1 (Units 1–5)

1 1 impression 2 astronomically 3 glamorous 4 variety 5 intriguing 6 confrontational 7 runner 8 supportive 9 Amazingly 10 trustworthy 11 rewarding 12 accidentally 13 responsibility 14 involvement 15 flexibility

2
1 I've just *crashed* into your car.
2 You ought *to* be strong …
3 … *which* was refurbished
4 just *in case* you need them.
5 *where* I used to work OR *that/which* I used to work *in* …
6 …everyone *had gone* home.
7 … you *will* have to improve …
8 … *to make* me look stupid/at making me look stupid in public.
9 … *would* that be a problem?
10 If I were *to* report you to the authorities OR If I *reported* you

3 We are all used to advertisers using (1) ~~the~~ sight and sound to grab our attention. Colourful logos are in (2) ~~an~~ evidence in our cities at every turn and TV commercials assail us with slogans and jingles whenever there is a break in the programmes. But (3) ~~the~~ most of us are not so aware that we can also be targeted through our sense of (4) ~~the~~ smell. The technique is not a new one. Supermarkets deliberately reheat their bread on the premises in the hope that the aroma will tempt shoppers to go to the bakery section on (5) ~~the~~ impulse. In the same way, many coffee shops like to grind (6) ~~a~~ coffee at the bar so that passers-by will be drawn in by the delicious smell. Now, however, a new device has been developed by a Japanese inventor, called an air cannon. This can single out a particular consumer and shoot a specified smell directly up their nose. With (7) ~~an~~ equipment like this, retail outlets are no longer confined to one or two long-lasting smells. The person by the fruit counter could get a whiff of citrus while another shopper looking for (8) ~~a~~ cleaning stuff could be targeted with a smell of (9) ~~the~~ beeswax or pine detergent. The result, no doubt, will be a greater number of purchases made without (10) ~~a~~ due consideration and, of course, more money for the stores.

4 1 anticlimax 2 misunderstood 3 outperformed 4 illogical 5 autobiography 6 underdeveloped 7 disobedient 8 irreversible 9 overspent 10 post-war

5 (wrong alternatives)
1 absolutely 2 very 3 utterly 4 very 5 wet 6 absolutely 7 extremely 8 devastated 9 skilful 10 terribly

6 1 do not have to write
2 is tested more rigorously
3 as often as she used
4 a common assumption that
5 chances are (that) he will forget
6 far fewer members (this year)
7 if/ should you happen to see him
8 regret not asking her
9 not nearly as/so difficult as
10 you (can) remember the combination
11 a dramatic rise in support
12 could have heard
13 allows us to foresee
14 I known (that) the pearls were (only)
15 are superior to

7 1 b) 2 a) 3 c) 4 b) 5 a) 6 b) 7 d) 8 b) 9 c) 10 c) 11 a) 12 b) 13 d) 14 b) 15 c) 16 b) 17 c) 18 d) 19 b) 20 d)

8 1 T 2 F 3 F 4 T 5 T 6 T 7 F 8 F 9 F 10 F

Unit 6 test

1 1 you didn't invite 2 I had followed 3 I could remember 4 would consent to it/would give his consent 5 I lived closer/nearer 6 you would stop staring 7 time you realised 8 I had let you 9 you told me (everything)/be told everything 10 we had not had an argument

2 1 so 2 it 3 to 4 there 5 so 6 ones 7 so 8 then 9 to 10 one

3 1 be 2 do 3 be 4 have (done) 5 do 6 have been

4 1 react *to* that 2 congratulated him *on* passing 3 concentrate *on* anything 4 comparing me *with* 5 She apologised *to* her mother 6 depends *on* him 7 boasting *about* it 8 have resulted *in* 9 complaining *about* his son 10 confide *in*?

5 1 ~~advise~~ *advice* 2 ~~principal~~ *principle* 3 ~~loose~~ *lose* 4 ~~stationery~~ *stationary* 5 ~~affect~~ *effect* 6 ~~practice~~ *practise*

6 1 c) 2 b) 3 a) 4 c) 5 d) 6 b) 7 a) 8 b)

Unit 7 test

1 1 I'm going to stay 2 is coming 3 I'll have left 4 I'll come 5 I'll be studying 6 will be 7 You'll break 8 will have finished 9 does your plane arrive? 10 we'll be enjoying

2 1 on the point of confessing 2 have no intention of putting 3 is due to open/begin/start 4 was (just) about to get 5 is in for a shock

3 1 to deceive 2 drawing 3 to make 4 to study 5 to revenge 6 to paint 7 to fool 8 to humiliate 9 producing 10 it to fall 11 serving 12 forging 13 to believe 14 them to let 15 him paint 16 us to re-examine 17 being 18 us think

4 1 snapping up 2 came across, was clearing out 3 was dashed off 4 brought about

5 1 b) 2 c) 3 c) 4 a) 5 d) 6 a) 7 b) 8 a) 9 d) 10 a) 11 a) 12 c)

Unit 8 test

1 1 had contributed 2 felt 3 had lost 4 didn't bring 5 might 6 would try 7 should 8 had managed

2 1 c) 2 b) 3 c) 4 a) 5 a) 6 a) 7 b) 8 b) 9 b) 10 c)

3 1 look down on 2 face up to 3 came up with 4 get round to 5 cut down on 6 run out of 7 came up against 8 put up with 9 break out of 10 get on with

4 1 c) 2 a) 3 c) 4 d) 5 b) 6 c) 7 a) 8 c) 9 d) 10 c) 11 c) 12 b) 13 b) 14 a) 15 d) 6 a)

5 1 disbelief 2 irrespective 3 independently 4 revelation 5 alienated 6 clarity

Unit 9 test

1 1 have always enjoyed 2 walked 3 had planned 4 beating 5 was doing/had done 6 had never suffered 7 passed 8 takes 9 was suffering 10 set 11 had hoped 12 to walk 13 visiting 14 am not planning

2 1 What we need to know is whether the transport strike is going ahead.
2 What they discovered was a fallen tree across the road.
3 What he did was advertise for a travelling companion on a website./What he advertised for on a website was a travelling companion.
4 What he said was that the town was virtually dead during the winter.
5 What happened was that they arrived at the waterfall but then they were caught in a storm.
6 What he enjoyed most was visiting the theme park.
7 What she did was fall asleep on the train and miss her stop.
8 What happened was that someone stole the money from his backpack.

3 1 spectacular 2 opening 3 of/on 4 stunningly 5 by 6 impossible 7 by 8 with 9 authenticity 10 from 11 with 12 resistance 13 to 14 to 15 inspiration 16 by 17 about 18 of

4 1 c) 2 d) 3 c) 4 b) 5 a) 6 c) 7 d) 8 b) 9 a) 10 a)

Progress test 2 (Units 6–10)

1 1 time you applied yourself 2 were on the point of reaching 3 is due to be released/for release 4 has difficulty (in) using 5 urged Mary to go 6 I had kept 7 surprises me is that 8 talked Nicholas out of taking 9 no point (in) standing around 10 that irritates me most about

2 1 ~~to laugh~~ laughing 2 ~~meet~~ meeting 3 ✓ 4 ✓ 5 ~~to touch~~ touch 6 ~~live~~ living 7 ~~to feel~~ feel 8 ~~criticising~~ criticise 9 ✓ 10 ~~to be~~ being

3 1 Relatively little research has been done into the ecosystem of the canopy.
2 Could you give me some advice on how to look after this plant?
3 There is not much evidence to suggest that wolves pose a serious threat to cattle.
4 ✓
5 The storm did a great deal of/a lot of damage to the new stadium.
6 The police found no proof that the animals on the farm were being mistreated.
7 ✓
8 I can't bear people who drop their litter on the street.
9 ✓
10 ✓

4 1 I have already made *it* clear … 2 ✓ 3 … we must take *it* seriously. 4 I find *it* appalling … 5 ✓ 6 We owe *it* to future generations … 7 ✓ 8 … that does not make *it* right. 9 ✓ 10 I take *it* that …

5 1 B 2 C 3 B 4 A 5 B 6 B 7 A 8 A 9 A 10 B 11 C 12 A 13 A 14 C

6 1 on 2 off/out 3 to 4 on 5 to 6 about 7 on 8 for 9 at 10 in

7 1 cut down 2 narrow down 3 water down 4 stir up 5 die down 6 track down 7 mount up 8 brush up

8 1 ~~controlled~~ checked 2 ~~loose~~ lose 3 ~~priceless~~ worthless 4 ~~effected~~ affected 5 ~~principal~~ principle 6 ~~logo~~ symbol 7 ~~rise~~ raise 8 ~~laying~~ lying 9 ~~display~~ present/put on 10 ~~hoard~~ accumulate/build up

9 1 b) 2 c) 3 d) 4 d) 5 b) 6 b) 7 a) 8 c) 9 a) 10 b) 11 b) 12 d)

10 1 C 2 A, B 3 B 4 A, B 5 A, B 6 C

Unit 11 test

1
1 can't have gone far
2 should have put on
3 might/may/could have been driving
4 ought to have told me
5 no way he could have known
6 sooner had they closed
7 only does he write
8 no time did the president admit
9 no circumstances should you reveal
10 had he finished his lecture
11 did people start to realise
12 do we remember exactly
13 did they realise
14 rarely do they spend money
15 should children be allowed to

2 1 b) 2 c) 3 b) 4 c) 5 a) 6 b) 7 d) 8 a) 9 c) 10 b) 11 b) 12 c) 13 b) 14 b) 15 a)

3 1 on 2 off 3 over/through 4 on 5 of 6 with 7 on 8 on 9 off 10 to

4 1 philosophically 2 enviable 3 unappreciated 4 helplessly 5 traumatic

5 1 remind 2 recall 3 memory 4 memorise 5 mind

Unit 12 test

1 1 to be thought of 2 were made to help 3 estimated to have attended 4 can be made/taken 5 is not felt to have done 6 is said to have died 7 have to/must be made 8 is thought to have been started 9 was to have been converted 10 was heard to say

2 1 was intended to be 2 was to become 3 would become 4 would attract 5 would have been 6 were aimed at improving 7 was planning to turn 8 was about to be struck 9 was to become 10 will be used

3 1 memorable 2 archeological 3 historical 4 speculation 5 reminiscences 6 disallowed 7 futuristic 8 visionary

4 1 c) 2 b) 3 b) 4 c) 5 c) 6 a) 7 c) 8 d) 9 b) 10 b) 11 d) 12 a)

5 1 course 2 will 3 trigger 4 plot 5 hold

6 Not possible:
1 b) 2 a) 3 a) 4 b) 5 c)

Unit 13 test

1
1 **Having** spent so much
2 answers **which/that** exceed
3 **Being** rather forgetful
4 ✓
5 Children **who** take
6 having **died** in the 1960s.
7 Having **been** brought up
8 **having** worked on the

2
1 have been much debated
2 is thought to have been written
3 was charged with wasting
4 have been dropped from
5 to being told
6 has now been widely replaced by/with
7 are being taken/adopted to prevent
8 should this policy be/if this policy should be

3 1 b) 2 a) 3 c) 4 c) 5 a) 6 c)

4 1 periodically 2 outselling 3 belatedly 4 overrated 5 truancy 6 comparatively 7 professionalism 8 availability 9 disruptive 10 underlying

5 1 a) 2 c) 3 a) 4 b) 5 d) 6 c) 7 b) 8 c) 9 a) 10 c) 11 b) 12 b) 13 d) 14 d) 15 c) 16 a) 17 b) 18 c)

Progress test 3 (Units 11–14)

1
1 could have been waiting
2 no circumstances should you
3 can't have been
4 had I lifted the receiver when
5 are believed to have been destroyed
6 sooner had they sat down than
7 was first struck by
8 having had parents who were (both)
9 was to have/should have been completed
10 could/may well have been
11 that can be done to help
12 having been tricked into (doing)
13 attention was paid to
14 take it for granted that
15 was about to be signed
16 (be sure to) take advantage of

2 1 ~~has~~ *have* 2 ~~you borrowed~~ *did you borrow* 3 ~~mustn't~~ *can't* 4 ~~I realised~~ *did I realise* 5 ~~would~~ *were planning to/were intending to/were going to* 6 ~~monitor~~ *be monitored* 7 ~~make~~ *be made* 8 ~~were crashing~~ *crashing/which were crashing/that were crashing*

3 1 beneficial 2 maximise 3 mobility 4 commercially 5 helplessly 6 objections 7 comparatively 8 worsened 9 hindsight 10 unwilling

4 1 received 2 disappointed 3 responsibility 4 accommodate 5 ✓ 6 useful 7 ✓ 8 embarrassing 9 recommend 10 Unfortunately 11 normally 12 although 13 Whether

5
1 All of the *books* that he wrote turned out to be best-sellers. (remove apostrophe)
2 People *say* that you should never judge a book by its cover. (remove comma)
3 It is probably the best of his *novels*; however, … OR It is probably the best of his *novels*. However, … (add full stop or semi-colon)
4 Helen White, whose first novel was published in *2002*, has just won an award. (add comma)
5 It was a long, *boring book*, written in a turgid style. (remove comma)
6 Despite popular *tradition*, Sherlock Holmes never says, 'Elementary, my dear Watson!' (add comma)
7 Both of the *students'* essays were outstanding pieces of work. (add apostrophe)
8 It is a carefully written account, but its general thesis is out of date. (remove apostrophe)
9 What really impressed me about the book was its attention to detail. (remove comma)
10 Shakespeare's plays can be divided into three *groups:* comedies, tragedies and histories. (add colon)
11 There are different theories about who wrote the sequel to the *poem*. (remove question mark and add full stop)
12 He keeps saying that the pen is his, but I'm sure that it's *ours*. (remove final apostrophe)
13 He wrote his first novel, *Hanging Man*, at the age of *17*, *which* was a remarkable achievement. (add comma)

6 1 b) 2 b) 3 c) 4 a) 5 d) 6 c) 7 b) 8 a) 9 c) 10 a) 11 d) 12 c) 13 a) 14 a) 15 d) 16 b) 17 b) 18 a) 19 d) 20 c)

7 1 F (I hour 10 minutes) 2 F 3 T 4 F 5 F 6 F (first part 180–220 words, second part 220–260 words) 7 T 8 F (it tests both) 9 F 10 F 11 T 12 T 13 T 14 F 15 T 16 T 17 F (one will just listen and assess) 18 F (you need to discuss or speculate about the situation) 19 T 20 F

Teacher's notes for photocopiable activities

1 Suffixes

Aim:

- **to give students practice in word building, focusing on suffixes**

Exam link

Paper 3 (Use of English), Part 3

Time

30 minutes

Preparation

Make copies of Worksheet 1 for one half of the class and copies of Worksheet 2 for the other half.

Procedure

1. Divide the class into two halves. Distribute Worksheet 1 to one half of the class and Worksheet 2 to the other half.
2. Ask students to work individually to complete the word diagrams in the first half of the Worksheet by inserting the words in the box in the correct place. Ask them to check this in pairs.
3. Students now work individually to try to complete sentences 1 to 7 with the correct form of the words in capitals. Emphasise that the answers to this exercise are not on the word diagrams they have just completed.
4. After they have completed sentences 1 to 7, each student swaps his/her Worksheet with a student in the other half of the class. Using the word diagrams on that Worksheet, they check their answers to the sentences.
5. Finally, go through the answers to both sets of sentences as a whole-class activity.

Options and alternatives

If students find this type of exercise difficult, you may choose to have the whole class working on the same sentences. In this case, students should work first on the word diagrams and then do the corresponding sentences, i.e. from the other Worksheet. You may also choose to do just one set of diagrams and sentences and save the others for another lesson.

ANSWERS

Worksheet 1

Imagine: imagination (noun), imaginative (adjective), imaginatively (adverb), unimaginative (negative adjective)
Disappoint: disappointment (noun), disappointing/ed (adjective)
Popular: popularity (noun), popularise (verb), unpopular (negative adjective)
Explain: explanation (noun), explanatory (adjective)
Innovate: innovation (noun), innovative (adjective)
Suspect: suspicion (noun), suspicious (adjective), suspiciously (adverb)
Commerce: commercial (adjective), commercialise (verb), commercialised (adjective)

1 variety **2** insistence **3** obediently **4** clarify **5** validate **6** advisable **7** likelihood

Worksheet 2

Obey: obedient (adjective), obediently (adverb), obedience (noun), disobey (negative verb)
Clear: clarity (noun), clarify (verb)
Likely: unlikely (negative adjective), likelihood (noun), unlikelihood (negative noun)
Vary: variety (noun), various (adjective)
Advise: advice (noun), advisable (adjective)
Valid: validity (noun), validate (verb), invalidate (negative verb)
Insist: insistence (noun), insistent (adjective)

1 disappointment **2** innovative **3** popularise **4** commercialised **5** suspiciously **6** unimaginative **7** explanatory

2A Compound adjective snap

Aim:

- **to review and extend students' knowledge of compound adjectives**

Exam link

Paper 3 (Use of English), Part 1
Paper 5 (speaking)

Time

30 minutes

Preparation
Make one copy of the adjectives and one copy of the participles for each pair of students and cut them into separate cards.

Procedure
1 Divide the students into pairs. Within each pair, give one student the adjective cards and the other student the participle cards.
2 Students lay out the cards face down in front of them. They then choose one of the cards in their pile and turn it over. If the words match to make a compound adjective, they should say 'snap'. The first person to say it collects that pair of cards. If the cards do not match they should be placed face down again, and another two cards turned over.
3 Students will know some of the compound adjectives from the Coursebook; others may be new. A student can challenge his/her partner's 'snap' by asking him/her what the adjective means. In cases of disagreement, they can refer to you. If necessary, you can impose the rule that after three false 'snaps', a player has to give all of his/her won cards to his/her partner.
4 Set a time limit of about 15 minutes for the game. At the end, the player in each pair who has collected the higher number of cards is the winner.
5 After the game, take the cards in and ask students to work in pairs to complete the gap-fill sentences 1–10, using compound adjectives that they remember from the game.

Options and alternatives
Instead of using the cards to play snap, students could work in groups and receive the adjective and particle cards in two separate envelopes. In this case, they simply work together to match up as many as they can before completing the gap-fill sentences.

ANSWERS

Possible pairs:
short-lived, short-tempered, short-sighted, short-staffed, hot-tempered, hot-headed, hot-blooded, quick-tempered, quick-witted, long-sighted, long-standing, long-suffering, long-winded, long-drawn-out, hard-earned, hard-headed, hard-pressed, level-headed, cold-blooded, self-satisfied, self-made, self-centred

Gap fill:
1 long-winded **2** short-lived **3** hard-headed (possibly: level-headed) **4** self-made
5 hard-pressed **6** short-staffed **7** long-drawn-out **8** long-standing **9** hard-earned
10 short-sighted

2B Advertising techniques

Aim:
- **to give practice in word building**

Exam link
Paper 3 (Use of English), Part 3 and Part 4

Time
40 minutes

Preparation
Make copies of Worksheet 2B Version 1 for one half of the class and copies of Version 2 for the other half.

Procedure
1 Ask students to brainstorm some techniques that advertisements use to attract our attention. Following this, write the term *subliminal advertising* on the board and ask if anyone knows what this means.
2 Tell students that they are going to read a text describing this advertising technique. Distribute Worksheet 2B Version 1 to half the class and Version 2 to the other half.
3 Students skim the text and then complete it, using the words at the bottom to form a word which fits into the same numbered space. They should note down their answers on another piece of paper.
4 When they have finished, they swap worksheets with a pair from the other half of the class and read the text to check their answers.

Options and alternatives
To make the exercise more challenging, do not give the words at the bottom of the text and ask students to do it as a gap fill. Then distribute the words and ask them to do the word-building exercise and compare their answers to this with their original answers.

ANSWERS
Version 1 answers are given on the Version 2 Worksheet and vice versa.

3 Personality types

Aim:
- **to introduce some vocabulary items to refer to different types of personalities**

Exam link
Paper 5 (speaking)

Time
20 minutes

Preparation
Make copies of Worksheet 1 for one half of the class and copies of Worksheet 2 for the other half, and a copy of the final definitions worksheet for each student.

Procedure

1 Distribute copies of Worksheet 1 to one half of the class (student A) and Worksheet 2 to the other (student B). Ask students to read through the mini-dialogues on their sheet. Point out that the second speaker in each case agrees with the first speaker by summarising his opinion of the personality with an idiomatic expression.
2 Ask students to look at the expressions in the box at the bottom of the sheet and try to work out what they mean. Some of these may be obscure unless students are familiar with the literary reference, as in *Peter Pan*, but others are guessable.
3 Students now work in pairs. One student (student A) reads the first part of mini-dialogue 1 only. His/her partner then completes the second speaker's agreeing phrase on Worksheet 2 by selecting the appropriate expression from the box. The first student should say immediately if the expression is correct or not. Student B then reads the first part of the mini-dialogue 2, and so on.
4 At the end of the activity, give each student the definition sheet and ask them to work in pairs to write each expression from the exercise next to the correct definition.

Options and alternatives
With a weaker group, you might choose to give them the expressions in the box and ask them to complete the definitions first, possibly using a dictionary, before going on to the pairwork. In this case, it would also be possible to turn the pairwork into a shorter, teacher-directed activity by just reading out some of the first speaker's descriptions to the class and asking them to name the personality type.

ANSWERS

Worksheet 1
2) high flyer **4)** a Scrooge **6)** tough cookie
8) Jekyll and Hyde **10)** busybody **12)** killjoy

Worksheet 2
1) dark horse **3)** Walter Mitty **5)** sponger
7) wannabe **9)** Peter Pan **11)** cold fish

Definitions
wannabe a person who wants to be famous for no good reason, or copies the behaviour of someone famous
Jekyll and Hyde a person who seems to have a split personality and is sometimes very pleasant and sometimes very unpleasant
high flyer a person who is ambitious and very successful in their work or studies
Scrooge a mean person
cold fish a person who seems unfriendly and without any strong feelings
dark horse a secretive person who does not tell other people much about their life
Walter Mitty a person who fantasises about leading an exciting, adventurous life when it is in fact quite ordinary
killjoy a person who complains about other people enjoying themselves or tries to spoil it for them
Peter Pan a person who behaves in a way which suggests that they are younger than they are
sponger a person who gets food or money from other people without offering to pay or doing anything in return
busybody a person who tries to interfere in what other people are doing
tough cookie a person who is clever but does not have much sympathy with other people's problems

4A Matching conditionals

Aim:

- **to provide further reinforcement of conditional structures and to introduce students to mixed conditionals**

Exam link
Paper 3 (Use of English), Parts 2 and 5

Time
20–30 minutes

Preparation
Make copies of the conditional sentences and cut them up so that students can match the two halves of each sentence.

Procedure

1 Divide students into pairs and give a copy of the cut-up conditional sentences to each pair.
2 Ask pairs to match the two halves to make correct conditional sentences. Tell them that there are also four halves which cannot match.
3 Check the answers either as a whole-class activity or by monitoring. Then draw students' attention to the 'mixed' second and third conditionals on the grid.
If + past perfect + *would* for an imaginary situation in the past with an imagined effect on the present.
e.g. *If small pox vaccinations had not been discovered, many children would still be dying of this disease.*
Had Crick and Watson not discovered the structure of DNA, there would be no biotechnology industry.

If + unreal past + *would have* for an imaginary situation in the present with an imagined effect on the past (often with the preposition *by*).
e.g. *If intelligent beings existed on other planets, then I am sure they would have contacted us by now.*
If that machine was ever going to be any good, you would have managed to make it work by now.

4 Now ask students to work in pairs to write 'if' clauses for the unmatched halves.
5 Finally they compare their sentences with another pair or read them out to the class.

Options and alternatives
You may choose not to introduce the mixed conditionals by leaving out the sentences in 3 above. In this case, the activity simply reinforces the structures introduced in the Coursebook.

> **ANSWERS**
> *The sentence halves are matched correctly prior to cutting.*

4B Idiom call my bluff

Aim:
- **to introduce students to some idioms further to those in the Coursebook**

Exam link
Paper 2 (Writing), Paper 3 (Use of English), Part 1

Time
20 minutes

Preparation
For one half of the class, make copies of Sheet 1 and Sheet 1 answers for each pair, and copies of Sheet 2 and Sheet 2 answers for each pair in the other half.

Procedure
1 Divide the class into two teams and within each team ask students to work in pairs. The pairs in team 1 receive Sheet 1 and pairs in team 2 receive Sheet 2.
2 The pairs discuss and decide which they think is the correct definition for the colour idiom. When they have decided, give them the answer sheet so that they can check their guess.
3 Each pair now joins up with a pair from the other team and asks them to discuss and guess the correct definition of the colour idioms.
4 Each pair now invents two false definitions of the two body parts idioms. They write these definitions plus the true one from the answer sheet in the three spaces under each idiom.
5 They now join up with the same pair as before from the other team and ask them to discuss and guess the true definitions.

Options and alternatives
In smaller groups, this can be done as a whole-class activity, with students working in two teams to guess and then present the three definitions to the other team. Instead of writing the definitions, three students can each present and explain one of the definitions to the other team so that the students must decide who is telling the truth. In this case, you could also introduce a rule that the guessing team is allowed to ask one question about the idiom and its use to each of the presenters. In this format, it is a good idea to write each idiom and the definition on the board after the guessing has taken place, to make sure that students do not remember the wrong definitions.

> **ANSWERS**
> *These are at the bottom of the Worksheet.*

5 Prefixes

Aim:
- **to review and extend students' knowledge of prefixes**

Exam link
Paper 3 (Use of English), Part 4

Time
20 minutes

Preparation
Make photocopies of Version 1 for one half of the class and Version 2 for the other half.

Procedure
1 Divide the students into pairs and give student A in each pair Version 1 and student B Version 2.
2 Give students one or two minutes to read through the sentences. They should then fill each gap with a word made by adding a prefix to one of the words in the box at the bottom of the sheet.
3 Student A now reads the first eight sentences to his/her partner, inserting what he/she thinks is the correct word. Student B checks against the complete sentences to see if the word is the same. At the end, he/she should tell his/her partner how many are correct but not which ones, and allow that number of second guesses. After this, he/she tells student A the correct answers and the final score.
4 Student B now reads sentences 9–16, inserting the correct words, while student A keeps the score and gives the answers at the end.

Options and alternatives

The exercise could be made easier by allowing students to tell their partner immediately whether the answer is right or wrong and allowing a second guess. Alternatively, all students could be given just the gapped sentences and the base words, and work in pairs to complete them. In this case, the exercise could be given for homework.

> **ANSWERS**
> *Version 1 answers are given on the Version 2 Worksheet, and vice versa.*

6 Birth order quiz

Aim:

- **to give practice in speaking, involving expressing and justifying opinions, making comparisons and hypothesising related to the topic of personality and family background**

Exam link

Paper 5 (speaking), Part 4

Time

30–40 minutes

Preparation

Make one copy of the quiz, Typical answers and Analysis per student.

Procedure

1 Begin by asking how many brothers or sisters the students have and establish how many eldest, middle, youngest and only children there are in the group.
2 If feasible, put the students into groups according to their birth order. Ask each group to think of three advantages and three disadvantages of their particular birth order. Then they feed their ideas back to the class.
3 Now ask students to work individually. They receive a copy of the quiz and work alone to choose their answers. Then they compare their answers with the typical ones for their birth order and read the analysis.
4 Ask students to work in pairs and tell their partner firstly the ways in which, according to the quiz, they are typical of people of their birth order and secondly the ways in which they are different.
5 Conduct a brief class discussion on how valid they feel the quiz is and how important birth order can be in determining personality.

Options and alternatives

It may not be possible to divide students into groups according to their birth order if numbers are very uneven. In this case, you can divide them into groups randomly and ask them to think of two possible advantages and two possible disadvantages for each, or simply discuss it briefly with the whole class before handing out the quiz.
Another possibility is to give out the quiz minus the title and with no initial discussion, so that students do not know what it is supposed to be testing. After they have completed it individually, tell them that one answer is supposed to be typical of eldest children, one of middle children and so on. They then discuss in groups which alternative they think corresponds to which birth order for each question, using their own answers and general opinions about eldest children, youngest children and so on to help them decide. Finally, they read the typical answers and analysis to see if they were right.

7 Reviews

Aims:

- **to introduce some vocabulary to talk about books and films**
- **to raise students' awareness of the organisation and information given in a review**

Exam link

Paper 2 (Writing), Part 2

Time

40 minutes

Preparation

Make one copy of Worksheet 1 for each student. Make copies of Worksheet 2A for one half of the class and of Worksheet 2B for the other and cut the extracts into separate strips.

Procedure

1 Distribute Worksheet 1 to the class. Ask them to work individually or in pairs to decide whether each of the sentences is from a review about a book or a film, or whether it could be either. Then ask them to decide if the comment is positive, negative or neutral in each case. You may wish to allow students to check some items in the dictionary before you go through the answers.
2 Divide the class into two halves. Give one half of the class the jumbled extracts for review A and the other the extracts for review B. Ask the class to work in pairs to put the extracts in the best order to create a book or film review. Tell them that there are two extracts that they should not include as they do not belong to their review.
3 Pairs now join up with a pair from the other half of the class. They decide together which two sentences have been swapped over in each case and where they should be placed in their review.
4 Ask students to choose two or three phrases from each review to record in their vocabulary notes.

Options and alternatives

You could shorten the activity by omitting step 1 and just doing the exercise with the jumbled reviews. You could also simplify the activity by not swapping any sentences over between the two reviews. In this case, students work in pairs to reconstruct either the complete book review or the complete film review and then simply ask a pair from the other group to read their review and comment on their order.

ANSWERS

Worksheet 1

	Book, film or both?	Positive, negative or neutral?
The special effects are amazing.	film	+
I couldn't put it down.	book	+
It is set in China during the time of the cultural revolution.	both	N
The ending is totally predictable.	both	–
It is a beautifully crafted piece of work.	both	+
The audience is quickly drawn into the story.	film	+
The clear and terse prose emphasises the banality of everyday life.	book	+
It holds up a mirror to life in a country village before the coming of the railways.	both	N
The plot turns on several well worn devices.	both	–
I found it totally absorbing.	both	+
It fails to live up to the promise of the first few chapters.	book	–
The writer has a fine ear for dialogue.	both	+
The ending falls very flat.	both	–
The beauty of the scenery is quite stunning.	film	+
I found it very cliché, not to mention sentimental.	both	–

Worksheet 2A

It is unusual for a writer to produce a best-selling novel while still in his teens. As a result, most novels which deal with the experiences of adolescence are written from a middle-aged point of view, with hindsight as it were.
But James Harding seems to be an exception. He began to write the first draft of his novel, Peak Times, at the age of 15. Four years later, it has now been published.
On the surface, the book holds up a mirror to street life in Glasgow. The action centres on a working-class family, their eldest son, Craig, and his relationship with the hostile, uncommunicative Jo.
When Jo runs away from home, Craig feels bound to follow her, which leads him into the frightening criminal underworld of Glasgow. The vision of the violent and totally amoral teenage gang culture is bleak.
Despite their misdeeds, both of the main characters come across as both vulnerable and, surprisingly, fundamentally good. I found myself turning the pages, hoping that their relationship would work out.
Even more surprisingly, Harding manages to give us an ending which offers a cautious hope for the couple.
The novel is written in a simple, economical style and the author has a fine ear for the local Glaswegian accent. At the same time, there are a number of subtle allusions to other literary works, most unexpectedly to Homer's Odyssey.
It is an impressive debut for someone so young and it will be interesting to see how his talent develops in the years to come

Worksheet 2B

Eric Norton is a gifted film maker and his latest work, Stars in Darkness, is close to being a masterpiece. Other works of his, such as My City, have dealt with the troubles of adolescence, but this work is probably his best yet.
The story is set in an unnamed seaside town in the South of England. The central character is Kevin, an only child of 14, whose life has been devastated by the death of his father.

We see how Kevin, bored and continually bullied at school, falls in with one of several gangs of local youths. Norton makes it clear how loyalty to the gang acts as a substitute for the family ties he has lost.
However, things turn much uglier with the arrival of 'Starman', a previous leader of the gang, recently released from prison.
After this new character turns up, we have a growing sense that events are building up to a catastrophe and when it happens it is felt to be both shocking and the inevitable culmination of everything that came before.
Brian Turnbull gives an impressive performance as Kevin and the other young members of the gang are equally strongly cast.
Grant Jackson's Starman, by contrast, seems a little overplayed and his immediate influence over even older members of the gang does not quite ring true.
Nevertheless this is a gripping and occasionally horrific film as well as being an acute commentary on the dangers and attractions of modern gang culture. It must definitely rank as one of the best to come out this year.

8A Risk transformations

Aims:

- **to review and extend the structures used with various reporting verbs**
- **to give practice in sentence transformations in preparation for Paper 3 (Use of English)**

Exam link
Paper 3 (Use of English), Part 5

Time
20 minutes

Preparation
Make one copy of the worksheet for each group of two or three students.

Procedure

1. Divide the class into groups of two or three students.
2. Give each group a copy of the worksheet and ask them to transform the sentences according to the instructions.
3. As well as doing the transformations, in each case they should decide how certain they are that their answer is correct and how many points they would like to risk on their answer. For example, if they are sure their answer is correct, they may risk 15 points, but if they are less sure they may choose to risk ten or just five.
4. When the students have finished, they swap papers.
5. Go through the answers with the class. The groups mark each others' papers. If the answer is correct, the team receive the number of points that they risked. If the answer is wrong, they lose those points.
6. The winning team is the team who has won the most points when all the sentences have been marked.

Options and alternatives
You could divide the class into two teams and ask each team to take it in turns to transform the sentences, which you write on the board. Before offering the transformation they should state how many points they are risking as in step 3 above.

ANSWERS

1. accused Gary of not doing his
2. suggested that the company should
3. claimed not to have taken/that he did not take
4. about the lack of enthusiasm among/from
5. congratulated Sophie on solving
6. would fall if they introduced
7. insisted on checking
8. for his dedication
9. to allow her to attend
10. complimented Julia on her knowledge

8B Compound nouns dice game

Aim:

- **to revise and extend students' knowledge of compound nouns formed with noun + preposition.**

Exam link
Paper 3 (Use of English), Parts 1 and 3

Time
20 minutes

Preparation
Make a copy of the verbs and prepositions table for each group of three or four students. You will also need a dice and possibly a dictionary for each group.

Procedure

1. Put the students into groups of three or four and give a copy of the verbs and prepositions table and a dice to each group.

2 Students take it in turns to throw the dice twice. The first throw gives the number of the preposition they must use and the second throw the number of the verb. For example, if a student throws a two and then a four, this gives the preposition *up* and the verb *set*.
3 The student must then give an example sentence for the noun formed from this verb and preposition, and explain the meaning if this is not clear from the example. If the other players agree it is correct, that player notes down the compound noun and gains a point. That particular compound cannot now be used again. If the other players disagree, they can appeal to you to decide. You may need to stress that for this activity, the combinations must be used as a compound noun, not a phrasal verb.
4 There are some verb–noun combinations where no word is possible. In this case, the turn passes immediately to the next player.
5 In some cases, there are two possibilities, e.g. *set-up* and *upset*. In this case, if one of the compounds has been used, the player can still gain a point by giving an example sentence for the other, unused compound.
6 After about 15 minutes of play, the winner is the student who has given correct sentences for the highest number of compounds.
7 To round the activity off, you may like to organise the students into different groups, so that they tell or teach each other the compounds that came up during their game.

Options and alternatives

If you would like students to extend their knowledge of compounds rather than just revising the ones presented in the Coursebook, you can introduce an additional rule into the game involving dictionary use. Give a dictionary to each group and tell them that each student is allowed to use the dictionary three times only during the game. If they feel that there may be a compound formed from a particular verb and preposition but are not sure, they may use the dictionary to check and, if they are right, gain the point. If you use this rule, you may want to conduct a brief feedback session afterwards where groups tell the class any new compounds they found, or again you could re-organise the groups so that they tell each other.

ANSWERS

Possible compounds:
outbreak, out-take, lookout, outlook, outset, outlet, outcome, break-up, uptake, set-up, upset, let-up, onset, setback, comeback, break-in, intake, inset, inlet, income, breakdown, let-down, comedown

9 Travel and transport idioms

Aim:

- **to introduce students to some idioms on the theme of transport**

Exam link

Paper 2 (Writing), Paper 3 (Use of English), Part 1

Time

25 minutes

Preparation

Make photocopies of Worksheet 1 for one half of the class and Worksheet 2 for the other half.

Procedure

1 Divide the class into two groups. Give copies of Worksheet 1 to the first group and Worksheet 2 to the second group. You may want to pre-teach one or two of the words in the idioms such as *rut*. Point out that the students do not need to use all the phrases in the boxes.
2 Ask the class to work either individually or in pairs to complete the sentences using the idioms in the table. They should start with whichever group of sentences is provided with clues. The clues should enable them to get most of these answers correct. When they have completed this set, then they move on to the second set and try to guess the answers from the remaining alternatives.
3 After they have completed the sentences, pair up students from the two different groups and ask them to read the clues to each other. Now that they have both sets of clues, they should be able to agree on the answers.
4 Finally, check the answers with the whole class.

Options and alternatives

Instead of putting students into pairs, they could stay in their teams to check the answers. In this case, ask the students who did not receive the clue to a particular sentence first if they can supply the correct idiom. The other team tells them if they are right or not and if they are wrong, prompts them by reading out the clue. Another alternative would be to give all students the sentences with the clues and simply ask them to work in pairs to complete them.

ANSWERS

Set A
1 took off 2 let off steam 3 missed the boat
4 soft landing 5 dead-end 6 all hands on deck

Set B
1 on the right track 2 one-track
3 go their separate ways
4 get them off the ground 5 at the helm
6 in a rut

10A Link words: Pickles and the world cup

Aim:

- **to review the meaning and use of various link words**

Exam link

Paper 2 (Writing), Part 2, Paper 3 (Use of English), Part 1

Time

20 minutes

Preparation

Make copies of Version 1 of the story 'Pickles and the world cup' for one half of the class and copies of Version 2 for the other.

Procedure

1. Tell students that they are going to read a story about a famous dog. You may like to pre-teach some of the vocabulary in the story such as *terrier*, *trophy* and *banquet*.
2. Divide the class into pairs. In each pair, one student receives Version 1 of the story and the other receives Version 2.
3. Tell the students that some of the information in their version of the story is incorrect. If it is incorrect, their partner has the correct version. Ask students to read their stories to each other. When they come across a difference, they should talk together to work out which of the two versions is correct at that point. To do this, students should pay close attention to the link word immediately before the two different versions, as it will only make sense with one of them. When they have worked out which is correct in each case, the student with the wrong version should cross it out and write his/her partner's version above.
4. Stress that students should read the story to each other and not just show it to their partner. You may like to tell them that one student should read the first and second paragraphs and his/her partner should read the third and fourth paragraphs.
5. Finally read the correct version to the class so that they can check their answers.

Options and alternatives

Instead of working in pairs at first, students could work individually to try to identify which parts of their narrative are incorrect and write a possible alternative on their sheet. They then work in pairs as above, comparing the corrections that they made with their partner's version.

ANSWER

Among England football fans, 1966 is remembered as the year in which the World Cup trophy was stolen. Fortunately, it was recovered in time for the game, not by a detective but by a black-and-white terrier called Pickles.

The famous Jules Rimet trophy was stolen from an exhibition hall on March 20 1966 despite the tight security surrounding it. The police immediately began an urgent investigation yet they completely failed to track down the missing cup. Later that week, Joe Mears, Chairman of the Football Association, received a call from a soldier called Edward Bletchley. Bletchley offered to return the cup for a sum of £15,000 as long as nothing was said to the police. Mears agreed to pay the ransom, but did not keep his promise to say nothing. Consequently, when Bletchley turned up at a secret location to collect the money, the police were ready to move in. He was arrested and charged with theft, although he later claimed to be only a middle man who would receive just £500 for his trouble.

Bletchley soon found himself in prison but the cup was still missing. Then, a few days later, a man called David Corbett was taking his dog Pickles for a walk when the dog dragged him over to a corner of the garden. Under the hedge was a parcel wrapped in newspaper. David's first thought was that it might be a bomb but when he pulled off the newspaper, there was the world cup. As a result of his find, Pickles became an instant celebrity. Because he had saved the world cup, he was allowed to attend the players' banquet and finish up the scraps. Moreover, David and Pickles went on to make a number of television appearances.

Although Pickles has been dead for many years, David still enjoys telling the story and is proud to have been the owner of one of the most famous dogs in history.

10B Commas and colons

Aim:

- **to revise the use of punctuation marks focusing especially on commas, colons and semicolons.**

Exam link

Paper 2 (Writing)

Time

20 minutes

Preparation

Make one copy of the worksheet for each student.

Procedure

1 The worksheet focuses on commas, colons and semicolons as these are the punctuation marks which students are most likely to be unsure how to use. Together with full stops, they are also the ones which they will most need for the CAE writing exam. For the section on commas, ask students to match the uses 1 to 4 with the example sentences a) to d). Then ask students to study the example sentences a) to d) for the colon and semicolon and complete the four rules about each punctuation mark.

2 Students then go on to the text. Tell them that there are some missing commas, colons and semicolons in the text. They should read the text carefully, checking it line by line. If a comma, colon or semicolon is missing in any line, they should insert it in the correct place. If the line is correct as it stands, they should put a tick at the end.

3 Finally go through the answers with the class, or give them a copy of the correct answer so that they can check their version.

Options and alternatives

If you wish to work more generally on punctuation, including full stops, you could dictate the first paragraph of the text to students with no punctuation and ask them to work in pairs to insert the full stops, commas, colons and semicolons.

ANSWERS

Commas

1 d) **2** b) **3** c) **4** a)

Colons and semicolons

1 colon (example sentence c)
2 semicolon (example sentence b)
3 colon (example sentence a)
4 semicolon (example sentence d)

1 ✓
2 colon after 'possibilities'
3 comma after first 'theory'
4 ✓
5 comma after 'stronger'
6 ✓
7 ✓
8 semi-colon after 'dinosaurs'
9 ✓
10 comma after 'bats'
11 semi-colon after 'true'
12 comma after 'similarity'
13 ✓
(Note: lines 7 and 9 could have commas added after 'tree to tree' and 'likely')

11A Emphatic inversion

Aims

- **to give freer practice of the structure for emphatic inversion introduced in the Coursebook**

Exam link

Paper 3 (English in Use) Part 5, Paper 5

Time

45 minutes

Preparation

Make copies of Part one of the worksheet for each student. Make a copy of the emphatic expressions for Part two for each group and cut them up into separate cards.

Procedure

1 Tell students that they are going to read a short speech made by a local politician. Distribute Part one of the worksheet and ask them to read the speech and insert the missing sentences into the most suitable gap.

2 Check the answers, drawing attention to the fact that the missing sentences all contain examples of emphatic inversion.

3 Tell students that they are going to work in groups of three and four to write their own political speech. Give them a choice of three or four subjects, e.g. traffic and parking problems, rubbish and litter problems, sporting facilities, facilities for recycling. It is best to choose subjects that are of significance locally so that students will have some ideas about what to say; however, avoid subjects which might be too sensitive.

4 Put the cards with the negative expressions into an envelope and ask a member of each group to draw out three. They must include each of these expressions in their speech, placed at the beginning of the sentences so that they use emphatic inversion.
5 After about 20 minutes ask one member of each group to read their speech to the class. You might like to allow one minute after each one for the class to ask the speaker any questions.
6 At the end, the class vote on which group wrote the best speech.

Options and alternatives

Instead of using the missing sentences exercise, you could turn the example speech into a listening exercise by reading it to the class. Before the first reading, ask a gist question such as *'What are the speaker's main criticisms of how the transport is run in his city?'* Read the speech again and ask the students to note the most important words in the four sentences with emphatic inversion. Then ask them to work in groups to reconstruct the four sentences.

ANSWERS

Part one

1 b) 2 d) 3 a) 4 c)

11B Eureka moments

Aim:

- **to give practice in word building**

Exam link

Paper 3 (Use of English), Part 3

Time

20 minutes

Preparation

Make copies of Worksheet 1 for one half of the class and copies of Worksheet 2 for the other half.

Procedure

1 Write the name *Archimedes* on the board and ask students what they know about him. Some will probably be able to tell you the famous story. Then ask if they know any other stories where the solution to a problem has come to someone in a flash like this, or if they have ever experienced this themselves.
2 Divide the class into two halves and distribute Worksheet 1 to one half and Worksheet 2 to the other. Students work in pairs or individually to complete the word-building exercise for their worksheet.
3 Each student now forms a pair with someone from the other group. They either read their texts to each other, while their partner checks their answers against his/her worksheet, or simply compare worksheets and mark their answers.

Options and alternatives

You could of course use either only Worksheet 1 or Worksheet 2 and use the text as a straight exam-style exercise. Alternatively, a strong group could work in pairs with one student doing the exercise orally and the other giving immediate feedback on his/her answers.

ANSWERS

Worksheet 1

1 loss 2 Overcome 3 miraculously
4 revelation 5 deduction 6 unconnected
7 lookout 8 unrelated

Worksheet 2

1 overflowed 2 realisation 3 insight
4 laborious 5 consideration 6 synthesising
7 unconsciously 8 inspiration

12 White elephants

Aim:

- **to give practice in fusing future in the past forms within a past narrative**

Exam link

Paper 2 (Writing), Paper 5 (Speaking)

Time

20 minutes

Preparation

Make one copy of Worksheet 1 for one half of the class and a copy of Worksheet 2 for the other half.

Procedure

1 Write the term *white elephant* on the board and ask the class if anyone knows the meaning. If not, explain that it means something useless or no longer needed but which has cost a lot of money. Tell the class that they are going to read about some famous white elephants.
2 Divide the class into two halves. Distribute Worksheet 1 to one half and Worksheet 2 to the other. Ask students to read the texts about the two famous white elephants. After reading the text they should complete three or four of the given sentences about each one, using future in the past forms.
3 Each student now forms a pair with someone from the other group. They tell their partner about each of the white elephants, using the information from the texts.

They should use the sentences they have written as part of their description.

4 Round off the activity by asking students if they can think of any other buildings or projects in their country or elsewhere which arguably have turned out to be white elephants.

Options and alternatives

You could of course shorten the activity by giving each half of the class the information about just one white elephant. With a strong group, you could then use one of the two remaining texts as a listening exercise by reading it to the class and then asking them to complete the sentences. They then compare their sentences in pairs.

13 Gapped sentences quiz

Aims:

- **to raise students' awareness of polysemy**
- **to give practice in completing an exam-style gapped sentences task**

Exam link

Paper 3 (Use of English), Part 4

Time

45 minutes

Preparation

Make copies of Worksheet 1 and the corresponding answers for one half of the class and copies of Worksheet 2 plus the corresponding answers for the other half.

Procedure

1 Divide the class into two groups. Give copies of Worksheet 1 to one group and Worksheet 2 to the other.

2 Ask the class to work in pairs to complete sentences 1 to 18 using the words in the box at the top. Tell them each word should be used three times, but, unlike in the exam, the groups of sentences are jumbled up.

3 When pairs have finished, give them a copy of the relevant answer sheet and ask them to check their answers. If there are any uses of the words that they do not understand they can check with you, or use a dictionary.

4 Now ask students to work in pairs and to rank the three sentences for each word according to how difficult they think the word is to guess. For example, on worksheet 1, students should rank sentences 1, 5 and 15 as they all require the word *rich*. The sentence they think is the most difficult of the three should be ranked number one and the easiest number three.

5 Students now join up with a pair from the other group. They should take it in turns to read the three sentences for one of the words to the other pairs. The other pair try to identify the word. Each pair should read the most difficult of the three sentences first and finish with the easiest. If the other pair identify the word correctly after the first sentence, they score three points. Two points are awarded if they identify it after the second sentence and only one if they need all three sentences. The pair are allowed to make notes, for example of collocations or prepositions in the sentences that are read to them, but they are only allowed one guess after each sentence.

6 At the end, the winners are the pair or pairs with the most points.

Options and alternatives

Instead of doing the quiz in groups of four, with a small class it could be done as a whole-class activity, with two teams reading the sentences across the class. In this case, all of the members of each team will have to agree on the ranking of the sentences. With a weaker class, you may wish to modify the rules and allow them more guesses, perhaps two after the first sentence in the group and just one after the others.

ANSWERS

The answers are given at the bottom of the Worksheet.

14 Spelling

Aims:

- **to focus on spelling rules and introduce some exceptions**
- **to give practice in proofreading for spelling mistakes**

Exam link

Paper 2 (Writing)

Time

20 minutes

Preparation

Make copies of Worksheets 1 and 2 for each student.

Procedure

1 Remind students of the spelling problems and rules given in the Coursebook such as '*i* before *e* except after *c*'. Then distribute Worksheet 1. Either go through the spelling rules with the class, or ask them to read them carefully themselves. Then students work individually to match each rule with the correct set of examples A to J. Allow them to compare in pairs before checking with the whole class. You might wish to add to rule 2 by telling the class that '*l*' is a special case and always doubles even in unstressed syllables, as in *travelled*; however, *traveled* is correct in US English.

2 Now distribute Worksheet 2. For each set of four words, students work individually to decide which is the one that is incorrectly spelt and write the correct spelling in the

'correction' column. They then decide whether the word that they have corrected is an example of one of the rules in Worksheet 1 or if it is an exception. They write either *exception* or *rule* in the third column.

3 After about ten minutes working on their own, students should compare and discuss their answers in pairs. Then go through the answers with the whole class. If the corrected word is an exception to the rule, ask if they know any other exceptions.

Options and alternatives

You could extend Worksheet 1 by asking students to supply further examples of words to illustrate each spelling rule. In this case, Worksheet 2 may become slightly easier, as they may well think of some of the words in the table.

This exercise, possibly without the 'exception or rule' column, could be repeated using the students' own spelling mistakes. Find about ten spelling mistakes from their written work and put each one with four correctly spelt words. Students again have to pick out and correct the misspelt word.

ANSWERS

Worksheet 1

1 C **2** E **3** J **4** F **5** H **6** B **7** G **8** A **9** I **10** D

Worksheet 2

correction	exception or rule?
1 beginning	rule
2 neighbour	rule
3 fortunately	rule
4 protein	exception
5 occurred	rule
6 disappear	rule
7 misspelt	rule
8 irresponsible	rule
9 studying	rule
10 successful	rule
11 courageous	exception
12 argument	exception

Notes on exceptions:

4 *seize* is also an exception.

11 In this case, the e is kept to retain the /dʒ/ sound of the *g*. This is also true in similar cases such as *advantageous* and *outrageous*.

12 The e is kept if there is another vowel before it. *Truly* is a similar exception.

Certificate in Advanced English Quiz p.182

ANSWERS

1 Yes, you have five minutes to do this at the end of the listening. In the reading, you should do this during the exam after you complete each section.
2 Pencil
3 This will be marked wrong, so you should rub one out.
4 For the listening, generally yes, although exceptions may be made for some difficult words. In the writing, many basic spelling errors will affect your mark but one or two minor spelling errors will be tolerated.
5 No
6 No, you can fail a paper and still pass the overall exam.
7 Usually about six weeks after the exam.
8 Four
9 75 minutes
10 It is a good idea to look at the questions first in Part 4. For the other three parts, read through the text first.
11 Two. The question in Part 1 is compulsory. In Part 2, you choose one from five possibilities (two of which refer to the set book).
12 No
13 Yes, but make sure the examiner can read your work.
14 No, because you won't have time. Write a plan for your answer and then write it out once.
15 Yes. Depending on the type of text, you might also use bullet points or subheadings for your paragraphs.
16 If your handwriting is difficult to read, the examiner may have to mark you down.
17 It can be, because it can make your answer easier to read and it is easier to correct things. However, you will need to indent your paragraphs if you do this.
18 If your answer is much too short you will be marked down. You will not automatically lose marks for an answer which is too long, but it is not a good idea to write an overlong answer as it may mean you have included things which are irrelevant. (If you are planning and checking your answer carefully, you won't have time anyway.)
19 Yes, as they are worth equal marks.
20 Five
21 This is not a good idea. You will only get the mark if both the answers are correct.
22 Yes
23 This will be marked wrong.
24 Four
25 Twice
26 You should write the words you hear if possible. You will not get extra credit for paraphrasing.
27 Yes
28 Yes
29 Examiners are trained to deal with this. Remember that if your partner does not listen to you or let you speak in Parts 3 and 4, he/she will be marked down, not you.
30 Find another way to say it.

Worksheet 1

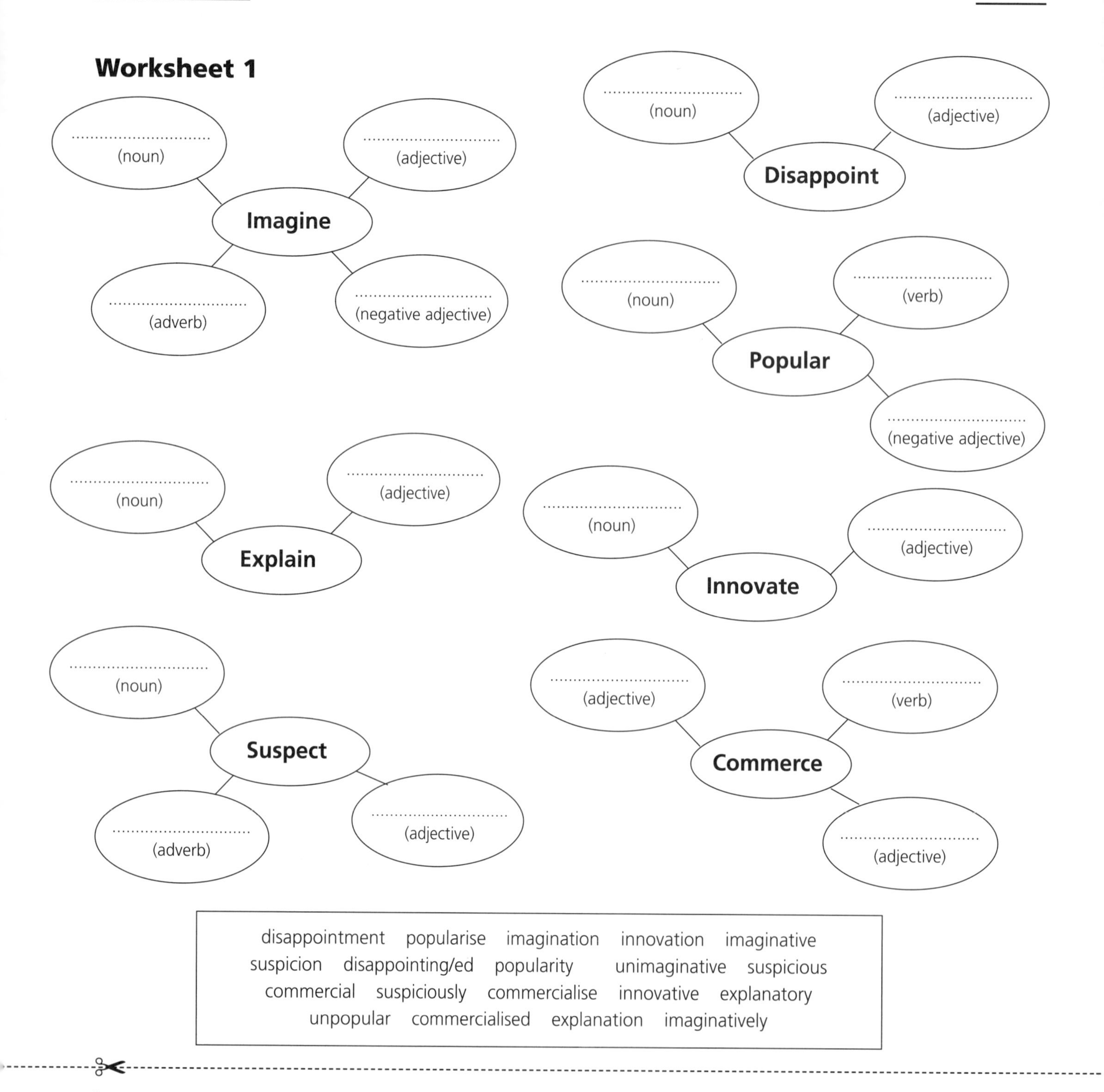

1 The station plays a wide .... of music, both popular and classical. **VARY**

2 The song was written at the .... of the record company, who wanted to appeal to a wider market. **INSIST**

3 After the piano lesson, he .... practised the new exercises for half an hour. **OBEY**

4 If a band composes a song together, they need to .... who owns the copyright before launching it. **CLEAR**

5 Although the research finds some exceptions, it does not .... the conclusion that downloading music from the Net decreases overall sales. **VALID**

6 It is .... to buy tickets for the concert well in advance. **ADVICE**

7 There is evidence to show that the use of personal stereos increases the .... of deafness in later life. **LIKELY**

Worksheet 2

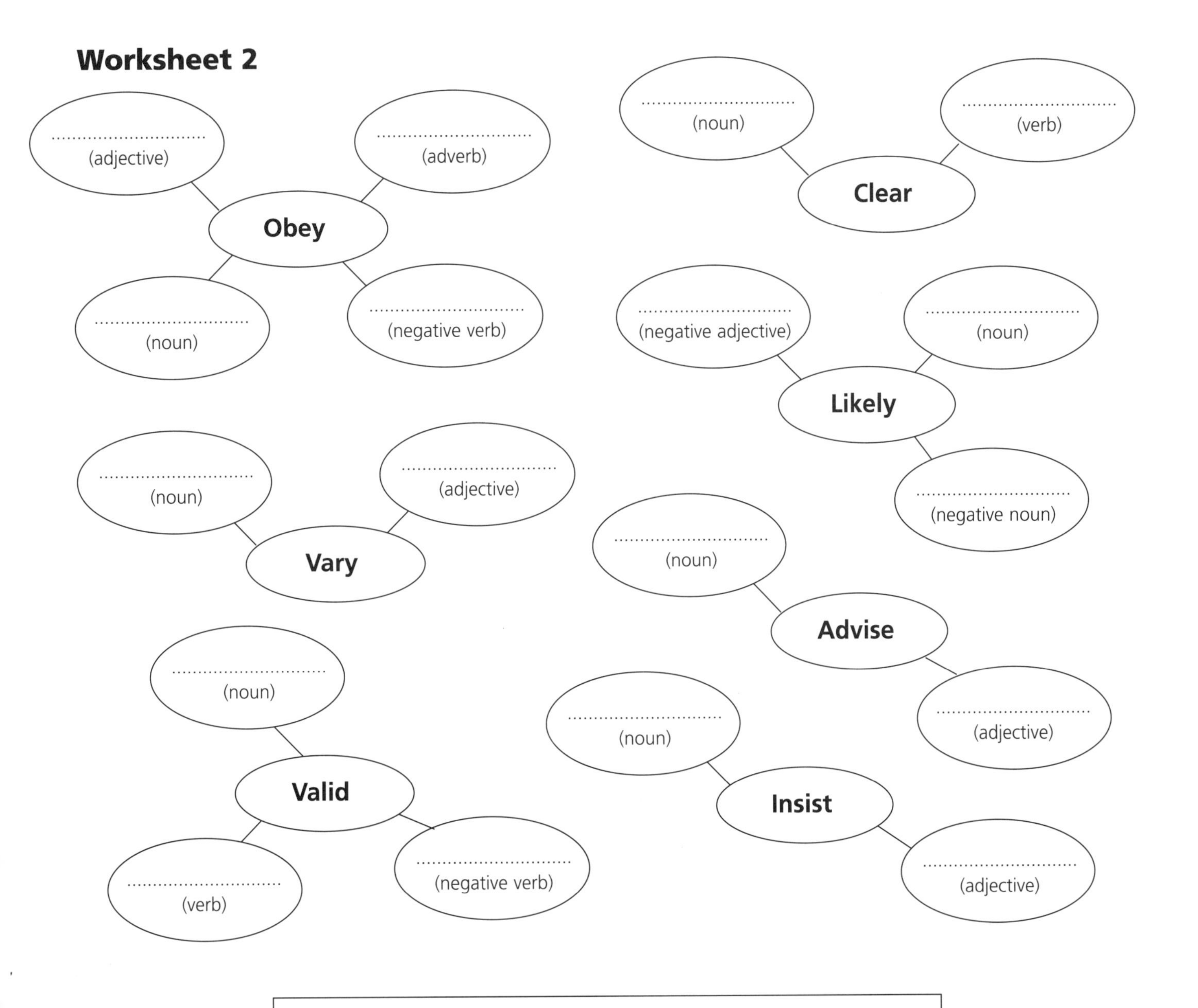

clarity advice obedience likelihood insistent clarify disobey insistence various validity obediently validate obedient unlikelihood advisable invalidate unlikely variety

---

1 We recommend that you book for the concert early to avoid .... **DISAPPOINT**

2 This is certainly one of the most .... productions I have ever seen. **INNOVATE**

3 Great musicians have helped to .... certain pieces of classical music. **POPULAR**

4 Music became very .... in the later twentieth century and punk rock was partly a reaction to this. **COMMERCE**

5 This new composition sounds .... like a reworking of one of their earlier pieces. **SUSPECT**

6 The new song was criticised for its banal lyrics and .... harmonies. **IMAGINE**

7 The book gives a clear account of the development of early music and has some good .... notes. **EXPLAIN**

SHORT	LONG
SHORT	LONG
SHORT	LONG
SHORT	HARD
HOT	HARD
HOT	HARD
HOT	LEVEL
QUICK	COLD
QUICK	SELF
LONG	SELF
LONG	SELF

LIVED	**WINDED**
TEMPERED	**DRAWN-OUT**
SIGHTED	**SUFFERING**
STAFFED	**EARNED**
HEADED	**HEADED**
TEMPERED	**PRESSED**
BLOODED	**HEADED**
TEMPERED	**BLOODED**
WITTED	**SATISFIED**
SIGHTED	**MADE**
STANDING	**CENTRED**

1 The finance director insisted on going through the budget in detail and talked for over half an hour. He was really very ....
2 His career as a restaurant owner was .... as the business went bankrupt within a year.
3 He was known as a .... businessman who never allowed emotion to cloud his judgements.
4 He started off selling pastries in the market but now he has over 300 retail outlets. He's a typical .... man.
5 I'm afraid we will be .... to find the money for a Christmas party this year. We really have so little to spare.
6 There have been problems with processing the overtime payments as the finance department are .... at the moment.
7 I am afraid pay negotiations have been rather .... as both sides refused to compromise.
8 We have dealt with that insurance company for years so we have an excellent and .... relationship.
9 Modern advertising methods persuade people to waste their .... money on unnecessary items.
10 To save money by making cuts in education is a very .... policy. It will benefit nobody in the long term.

Version 1

SUBLIMINAL ADVERTISING

A subliminal message is a message planted in another form of media which bypasses the usual channels of (1) .... and is only registered (2) ....

In 1957, James Vicary claimed that he had (3) .... used this technique in advertising. During a film, he had (4) .... flashed the slogans 'Drink Coca Cola™' and 'Eat popcorn' for an (5) .... 1/3000 of a second. The result was that the message had been implanted in the brains of the (6) .... audience and sales of cola and popcorn at that cinema had increased (7) ....

The immediate response to this claim was a public outcry and a number of governments outlawed the use of such techniques. However, doubts about the claim soon set in. Later studies were unsuccessful in their attempts to replicate Vicary's results and indeed seemed to suggest that subliminal advertising was totally ineffective. In 1962, Vicary finally confessed that he had falsified the data of the original experiment. There is in fact no decisive evidence to suggest that audiences can be manipulated by this technique. Even if it has an effect it is probably only like a momentary glance at a billboard.

1 PERCEIVE **2** CONSCIOUS **3** SUCCEED **4** CONTINUE
5 PERCEIVE **6** SUSPECT **7** DRAMA

Version 2

SUBLIMINAL ADVERTISING

A subliminal message is a message planted in another form of media which bypasses the usual channels of perception and is only registered unconsciously.

In 1957, James Vicary claimed that he had successfully used this technique in advertising. During a film, he had continually flashed the slogans 'Drink Coca Cola™' and 'Eat popcorn' for an imperceptible 1/3000 of a second. The result was that the message had been implanted in the brains of the unsuspecting audience and sales of cola and popcorn at that cinema had increased dramatically.

The immediate (8) .... to this claim was a public outcry and a number of governments (9) .... the use of such techniques. However, doubts about the claim soon set in. Later studies were (10) .... in their attempts to replicate Vicary's results and indeed seemed to suggest that subliminal advertising was totally (11) .... In 1962, Vicary finally confessed that he had (12) .... the data of the original experiment. There is in fact no (13) .... evidence to suggest that audiences can be manipulated by this technique. Even if it has an effect it is probably only like a (14) .... glance at a billboard.

8 RESPOND **9** LAW **10** SUCCEED **11** EFFECT
12 FALSE **13** DECIDE **14** MOMENT

Worksheet 1

1 'I just think there might be sides to her that we don't know anything about. No one seems to know anything about her life outside work.'
'Yes, she's a bit of a dark horse.'

2 'Yes. He's definitely a ………… .'

3 'He was telling me the most amazing stories about his career as a professional stuntman. I don't believe a word of it.'
'Yes, he's well known as a bit of a Walter Mitty.'

4 'He's such a ………… .'

5 'He came out with us last night but he didn't contribute anything to the cost of the meal or even buy anyone else a drink. Then he even asked if he could share my taxi home.'
'What a sponger!'

6 'A bit of a …………, then.'

7 'She thinks she's going to be famous but as far as I can see she's got no talent for anything.'
'Don't take her too seriously. She's just a wannabe.'

8 'I know. It's ………… all over again.'

9 'He wants to stay young, free and single even though he's in his forties. He looks about half his age as well.'
'Yes, he's a Peter Pan really, isn't he?'

11 'Yes, she's a bit of a ………… .'

11 'He never seems to smile at anyone or show any enthusiasm for anything. He's really not my idea of fun.'
'A bit of a cold fish, then.'

12 'What a …………!'

tough cookie	Scrooge	high flyer	killjoy	Jekyll and Hyde	busybody

Worksheet 2

1 'Yes, she's a bit of a ....'
2 'I'm sure he's got his eye on the top job and with the amount of work he does, he'll probably get it too.' 'Yes. He's definitely a high flyer.'
3 'Yes, he's well known as a bit of a ....'
4 'He wouldn't give anything at all towards my sister's leaving present. He just said he couldn't afford it. Can you believe that?' 'He's such a Scrooge.'
5 'What a ....!'
6 'Some people say she's unfeeling but she's really good at running the business and making it profitable. She'd never let her emotions cloud her judgement.' 'A bit of a tough cookie, then.'
7 'Don't take her too seriously. She's just a ....'
8 'I can't make him out really. Sometimes he seems so kind and helpful and yet at other times he'll be really unpleasant and aggressive.' 'I know. It's Jekyll and Hyde all over again.'
9 'Yes, he's a .... really, isn't he?'
10 'She's always trying to interfere in other people's lives. I nearly told her today that what I do in my spare time has nothing to do with her.' 'Yes, she's a bit of a busybody.'
11 'A bit of a ...., then.'
12 'I can't stand our neighbour. Every time we have a party he complains about the noise. He even complains about children playing in the park.' 'What a killjoy!'

Peter Pan
cold fish
Walter Mitty
wannabe
sponger
dark horse

Definitions

	a person who wants to be famous for no good reason, or copies the behaviour of someone famous
	a person who seems to have a split personality and is sometimes very pleasant and sometimes very unpleasant
	a person who is ambitious and very successful in their work or studies
	a mean person
	a person who seems unfriendly and without any strong feelings
	a secretive person who does not tell other people much about their life
	a person who fantasises about leading an exciting, adventurous life when it is in fact quite ordinary
	a person who complains about other people enjoying themselves or tries to spoil it for them
	a person who behaves in a way which suggests that they are younger than they are
	a person who gets food or money from other people without offering to pay or doing anything in return
	a person who tries to interfere in what other people are doing
	a person who is clever but does not have much sympathy with other people's problems

If we invested in this new software …	… we would have to train staff how to use it.
If smallpox vaccinations had not been discovered …	… many children would still be dying of this disease.
Unless the sales of digital cameras start to pick up …	… we will have to consider making cuts in our workforce.
If the virus enters your computer …	… it will destroy all the data on your hard disc.
If intelligent beings existed on other planets …	… then I am sure they would have managed to contact us by now.
If we had not had such good technical support …	… the show would have been a disaster.
Had Crick and Watson not discovered the structure of DNA …	… there would be no biotechnology industry.
Unless this device can be patented …	… it will probably be copied by many rival companies.
If planes were to fly at a different altitude …	… they could reduce their contribution to global warming.
If that machine was ever going to be any good …	… you would have managed to make it work by now.
If our sun were to turn into a black hole …	… our solar system would immediately be sucked in.
If you happen to pass the computer store …	… buy me another memory stick, will you?
Supposing you took your product to the exhibition …	… do you think many people would be interested?
Were they to install a robot doctor on the spacecraft …	… the astronauts could monitor their health daily.
… the show would fail on its first night.	… they will not look anything like us.
… the disease could be eradicated from the world.	… the astronauts would not have survived the trip.

Sheet 1

TO PAINT THE TOWN RED

1 to enjoy yourself in a lot of bars and public places

2 to persuade someone to adopt left-wing political views

3 to steal everything you can from a person or institution

TO BE ALL FINGERS AND THUMBS

1 ....

2 ....

3 ....

TO BE WET BEHIND THE EARS

1 ....

2 ....

3 ....

Sheet 2

TO BE IN THE PINK

1 to have just enough money to live on

2 to be newly married

3 to feel very fit and healthy

TO HAVE YOUR TONGUE IN YOUR CHEEK

1 ....

2 ....

3 ....

TO PUT YOUR FOOT DOWN

1 ....

2 ....

3 ....

Sheet 1 answers

To paint the town red: definition 1

To be all fingers and thumbs: to be clumsy

To be wet behind the ears: to be young and inexperienced

Sheet 2 answers

To be in the pink: definition 3

To have your tongue in your cheek: to say something which is not meant to be taken seriously

To put your foot down: to be very strict and insist on something

Version 1

1 Some people think that sports stars are grossly .... when so many families live in poverty.

2 Several football fans were arrested for their .... conduct.

3 He is a very .... player. Sometimes he plays really well and at other times he's no good at all.

4 Don't .... Sheffield United! They are a much better team than you think.

5 Alex Morgan suffered a knee injury earlier this season and it still causes him ....

6 That must be one of the most .... matches this season! Sharapova's performance was incredible.

7 Mr Bennet was sacked as team manager but they .... him after a year.

8 Unfortunately, their hopes of winning the cup are looking more and more ....

9 He was disqualified from taking part in the race after failing the drugs test.

10 The match was abandoned because of the stormy weather and they still have not fixed a date for the replay.

11 He is one of the best players we have ever had, but he's still not infallible.

12 The national football federation has been accused of financial mismanagement.

13 The tennis player miscalculated his serve and sent the ball right out of the court.

14 She is a good athlete but I think her plan to go in for the Olympic Games is rather overambitious.

15 It was almost unbearable to watch the team play so badly after all the coaching we had done.

16 It was irresponsible of the sports channel to give interviews to a group of football hooligans.

forgettable	estimate	paid	consistent	orderly	comfort	realistic	instate

Version 2

1 Some people think that sports stars are grossly overpaid when so many families live in poverty.

2 Several football fans were arrested for their disorderly conduct.

3 He is a very inconsistent player. Sometimes he plays really well and at other times he's no good at all.

4 Don't underestimate Sheffield United! They are a much better team than you think.

5 Alex Morgan suffered a knee injury earlier this season and it still causes him discomfort.

6 That must be one of the most unforgettable matches this season! Sharapova's performance was incredible.

7 Mr Bennet was sacked as team manager but they reinstated him after a year.

8 Unfortunately, their hopes of winning the cup are looking more and more unrealistic.

9 He was ………… from taking part in the race after failing the drugs test.

10 The match was abandoned because of the stormy weather and they still have not fixed a date for the ………… .

11 He is one of the best players we have ever had, but he's still not ………… .

12 The national football federation has been accused of financial ………… .

13 The tennis player ………… his serve and sent the ball right out of the court.

14 She is a good athlete but I think her plan to go in for the Olympic Games is rather ………… .

15 It was almost ………… to watch the team play so badly after all the coaching we had done.

16 It was ………… of the sports channel to give interviews to a group of football hooligans.

bearable	calculated	fallible	qualified	responsible	management	ambitious	play

DOES YOUR BIRTH ORDER MATCH YOUR PERSONALITY?

1 How would you describe your social life?
- a I have a small number of like-minded friends who I trust.
- b I have many friends and get on with many different types of people.
- c I have lots of friends but very few close ones.
- d I am something of a loner with just one or two close friends.

2 How do you normally react if someone criticises you harshly?
- a I tell myself that they don't really know me.
- b I shrug it off and laugh about it with friends afterwards.
- c I say very little but feel upset for some time afterwards.
- d I tell them not to speak to me like that.

3 What age are your friends?
- a They are often older than me.
- b They are usually the same age as me or younger.
- c They may be older or younger than me. It doesn't matter.

4 What kind of celebrity would you like to be?
- a President of your country.
- b Someone who champions a good cause.
- c A writer or philosopher.
- d A rock singer.

5 What kind of pet would you prefer?
- a A loyal obedient pet, like a dog.
- b A more independent pet, like a cat.
- c A lovable, crazy pet.
- d An unusual pet.

6 Which is your biggest strength at work?
- a Your ability to deal with and manage difficult people.
- b Your leadership and your ability to organise others.
- c Your ability to set yourself goals and achieve them.
- d Your creativity and your ability to come up with innovative ideas.

7 At work, which do you find most difficult?
- a Meeting deadlines.
- b Admitting that you need help.
- c Motivating yourself to work alone.
- d Delegating.

8 Your boss is asking you to do extra tasks that are not part of your job at all. What do you do?
- a Do them with good grace. After all, the fact that s/he is asking you shows that they trust you.
- b Do them, but seethe about it to yourself.
- c Only do them if you have time. You probably won't!
- d Explain how you feel and try to reach a compromise.

9 Where do you prefer to go on holiday?
- a A place you have visited before and know you will like.
- b Somewhere quiet where you can escape the rat race.
- c Somewhere new that you can have fun exploring.
- d Somewhere you can let your hair down and enjoy yourself.

10 During your childhood, how do you think your parents treated you?
- a They had very high expectations of me.
- b They did not give me enough attention.
- c They let me get away with almost anything.
- d They gave me too much attention.

TYPICAL ANSWERS

Eldest children

1 a 2 d 3 b 4 a 5 b 6 b 7 d 8 a 9 a 10a

Middle children

1 b 2 a 3 c 4 b 5 a 6 a 7 c 8 d 9 c 10 b

Youngest children

1 c 2 b 3 c 4 d 5c 6d 7 a 8 c 9d 10 c

Only children

1 d 2 c 3 a 4c 5 d 6 c 7 b 8 b 9 b 10 d

ANALYSIS

Eldest children

Eldest children typically receive high expectations from their parents (question 10). As a result, they often become hard working and perfectionist. This can mean that they are not good at delegating, as they do not trust others to do something as well as they can (question 7). They are good at organising, and can occasionally be authoritarian (question 6). They are also rather conservative. They like what they are used to (question 9) and dislike surprises or innovation.

Middle children

Middle children frequently feel that they have missed out on a position of power and responsibility (question 10). Typically, they are diplomatic, good at dealing with others and relating to a wide range of people (questions 1 and 6). They often find it difficult to feel motivated without others to drive them (question 7). They tend to identify with causes (question 4) and often feel strongly about the injustices of the world.

Youngest children

Youngest children tend to receive less discipline and feel under less pressure to achieve than their older siblings (question 10). They are often charming and outgoing but can also be rather superficial (question 1). They are often innovative and creative (question 6) but may lack the willpower to persevere or find it difficult to meet deadlines (question 7). They are less happy with responsibility than other groups but often crave excitement or fame (question 4).

Only children

Only children are rather similar to eldest children in that they receive high expectations and a great deal of attention from parents (question 10). They too are often hard working and perfectionist. They are good at setting themselves goals and working independently for long periods (question 7). On the negative side, they are not always good at communicating their feelings to others (question 8). As they will have spent more time in an environment without other children, they may relate best to people older than themselves (question 3).

Worksheet 1

	Book, film or both?	Positive, negative or neutral?
The special effects are amazing.		
I couldn't put it down.		
It is set in China during the time of the Cultural Revolution.		
The ending is totally predictable.		
It is a beautifully crafted piece of work.		
The audience is quickly drawn into the story.		
The clear and terse prose emphasises the banality of everyday life.		
It holds up a mirror to life in a country village before the coming of the railways.		
The plot turns on several well worn devices.		
I found it totally absorbing.		
It fails to live up to the promise of the first few chapters.		
The writer has a fine ear for dialogue.		
The ending falls very flat.		
The beauty of the scenery is quite stunning.		
I found it very cliché, not to mention sentimental.		

Worksheet 2A

It is unusual for a writer to produce a best-selling novel while still in his teens. As a result, most novels which deal with the experiences of adolescence are written from a middle-aged point of view, with hindsight as it were.

But James Harding seems to be an exception. He began to write the first draft of his novel, Peak Times, at the age of 15. Four years later, it has now been published.

On the surface, the book holds up a mirror to street life in Glasgow. The action centres on a working-class family, their eldest son, Craig, and his relationship with the hostile, uncommunicative Jo.

When Jo runs away from home, Craig feels bound to follow her, which leads him into the frightening criminal underworld of Glasgow. The vision of the violent and totally amoral teenage gang culture is bleak.

After this new character turns up, we have a growing sense that events are building up to a catastrophe and when it happens it is felt to be both shocking and the inevitable culmination of everything that came before.

Even more surprisingly, Harding manages to give us an ending which offers a cautious hope for the couple.

The novel is written in a simple, economical style and the author has a fine ear for the local Glaswegian accent. At the same time, there are a number of subtle allusions to other literary works, most unexpectedly to Homer's Odyssey.

Nevertheless this is a gripping and occasionally horrific film as well as being an acute commentary on the dangers and attractions of modern gang culture. It must definitely rank as one of the best to come out this year.

Worksheet 2B

Eric Norton is a gifted film maker and his latest work, Stars in Darkness, is close to being a masterpiece. Other works of his, such as My City, have dealt with the troubles of adolescence, but this work is probably his best yet.

The story is set in an unnamed seaside town in the south of England. The central character is Kevin, an only child of 14, whose life has been devastated by the death of his father.

We see how Kevin, bored and continually bullied at school, falls in with one of several gangs of local youths. Norton makes it clear how loyalty to the gang acts as a substitute for the family ties he has lost.

However, things turn much uglier with the arrival of 'Starman', a previous leader of the gang, recently released from prison.

Despite their misdeeds, both of the main characters come across as both vulnerable and, surprisingly, fundamentally good. I found myself turning the pages, hoping that their relationship would work out.

Brian Turnbull gives an impressive performance as Kevin and the other young members of the gang are equally strongly cast.

Grant Jackson's Starman, by contrast, seems a little overplayed and his immediate influence over even older members of the gang does not quite ring true.

It is an impressive debut for someone so young and it will be interesting to see how his talent develops in the years to come.

For questions 1 to 10 complete the second sentence so that it has a similar meaning to the first sentence, using the word given. Do not change the word given. You must use between three and six words including the word given.

		5	10	15	Marks
1	Ben said, 'Gary, you're not doing your fair share of the work'. **accused** Ben .... fair share of the work.				
2	Mr Green said, 'Why doesn't the company advertise on the Internet?' **suggested** Mr Green .... advertise on the Internet.				
3	Maurice said, 'I didn't take any days off sick last year'. **claimed** Maurice .... any days off sick last year.				
4	Mr Walker said 'The staff really aren't very enthusiastic'. **lack** Mr Walker complained .... the staff.				
5	Mrs Mason said, 'Well done, Sophie – I think you've solved a major problem there'. **congratulated** Mrs Mason .... a major problem.				
6	Clive said, 'I think introducing flexible hours will cause their productivity to fall'. **if** Clive said he thought that their productivity .... flexible hours.				
7	Mr Cole said, 'I am certainly going to check the figures myself'. **insisted** Mr Cole .... the figures himself.				
8	Warren said, 'Alan you've been really dedicated to the project'. **his** Warren praised Alan .... to the project.				
9	Anne said, 'Geoff, please let me attend the union meeting'. **allow** Anne begged Geoff .... the union meeting.				
10	Mr Rayner said, 'Julia, you know so much about market research'. **complimented** Mr Rayner .... of market research.				

Verbs

Prepositions

	❶ OUT	❷ UP	❸ ON	❹ BACK	❺ IN	❻ DOWN
❶ BREAK						
❷ TAKE						
❸ LOOK						
❹ SET						
❺ LET						
❻ COME						

Worksheet 1

took a nosedive	soft landing	missed the boat
dead-end	on the rocks	let off steam
go their separate ways	in a rut	at the helm
one-track	on the right track	get them off the ground
took off	all hands on deck	

Set A

1 His career really .... after his first television appearance.
(His career is like a plane which has now started to fly successfully.)

2 It's good that we live so near the park. The children need a nice big open space where they can run around and ....
(The children are like trains whose engines may become overheated if they are not active.)

3 If you don't take this opportunity now, you may find you've ....
(Taking this opportunity would be like beginning a journey by sea.)

4 Despite all the predictions about a stock market crash, the situation now looks more hopeful and I'm confident we will have a ....
(The stock market is like a plane. Instead of the flight ending in a crash, it will now end safely.)

5 There are no opportunities for promotion and no training offered. It's a real .... job.
(The job is like a road but it does not lead anywhere.)

6 No hotel staff are allowed to go on holiday during the summer as we really need .... at that time.
(Working for the hotel is like working on a ship.)

Set B

1 I had a talk with the chief this morning and he definitely thinks the investigation is ....

2 Whenever you try and have a conversation with him, he starts talking about his promotion. He's got a real .... mind.

3 They were business partners for five years before they decided to ....

4 Unless all the team are fully committed to the projects, you won't be able to ....

5 After Mr Grove's poor performance in parliament yesterday, the democrats must be wondering if they have the right leader ....

6 He's frustrated with his job and feels he's going nowhere. He's really stuck ....

Worksheet 2

took a nosedive	soft landing	off the rails
missed the boat	dead-end	on the rocks
let off steam	go their separate ways	back seat driver
in a rut	at the helm	one-track
on the right track	get them off the ground	took off
all hands on deck		

Set A

1 His career really .... after he joined the new marketing department.

2 It's good that we live so near the park. The children need a nice big open space where they can run around and ....

3 If you don't take this opportunity now, you may find you've ....

4 Despite all the predictions about a stock market crash, the situation now looks more hopeful and I'm confident we will have a ....

5 There are no opportunities for promotion and no training offered. It's a real .... job.

6 No hotel staff are allowed to go on holiday during the summer as we really need .... at that time.

Set B

1 I had a talk with the chief this morning and he definitely thinks the investigation is ....
(The investigation is like a train taking us where we want to go.)

2 Whenever you try and have a conversation with him, he starts talking about his promotion. He's got a real .... mind.
(His mind is like a train which can only travel in one direction.)

3 They were business partners for five years before they decided to ....
(They were like two people walking down the same road but now they are taking different roads.)

4 Unless all the team are fully committed to the new projects, you won't be able to ....
(The projects are like planes which need to fly successfully.)

5 After Mr Grove's poor performance in parliament yesterday, the democrats must be wondering if they have the right leader ....
(The political party is like a ship that Mr Grove is steering.)

6 He's frustrated with his job and feels he's going nowhere. He's really stuck ....
(The job is like a journey but he cannot go any further because he is caught in a hole in the road.)

Version 1

PICKLES AND THE WORLD CUP

Among England football fans, 1966 is remembered as the year in which the world cup trophy was stolen. Fortunately, it was recovered in time for the game, not by a detective but by a black-and-white terrier called Pickles.

The famous Jules Rimet trophy was stolen from an exhibition hall on 20 March 1966 **despite** the poor security arrangements. The police immediately began an urgent investigation **yet** they completely failed to track down the missing cup. Later that week, Joe Mears, Chairman of the Football Association, received a call from a soldier called Edward Bletchley. Bletchley offered to return the cup for a sum of £15,000 as long as nothing was said to the police. Mears agreed to pay the ransom, **but** did not keep his promise to say nothing. **Consequently**, when Bletchley turned up at a secret location to collect the money, the police were ready to move in. He was arrested and charged with theft, **although** he later claimed to have devised the whole plot to steal the trophy himself.

Bletchley soon found himself in prison but the cup was still missing. Then, a few days later, a man called David Corbett was taking his dog Pickles for a walk when the dog dragged him over to a corner of the garden. Under the hedge was a parcel wrapped in newspaper. David thought immediately that it might be the missing trophy **but** when he pulled off the newspaper, there was the world cup. **As a result of** his find, hardly any one realised that the dog was responsible. **Because** he had saved the world cup, the Football Association did not even mention him at their banquet. **Moreover**, David soon dropped from the public eye as well.

Although Pickles has been dead for many years, David still enjoys telling the story and is proud to have been the owner of one of the most famous dogs in history.

Version 2

PICKLES AND THE WORLD CUP

Among England football fans, 1966 is remembered as the year in which the world cup trophy was stolen. Fortunately, it was recovered in time for the game, not by a detective but by a black-and-white terrier called Pickles.

The famous Jules Rimet trophy was stolen from an exhibition hall on 20 March 1966 **despite** the tight security surrounding it. The police immediately began an urgent investigation **yet** came very close to finding the missing cup. Later that week, Joe Mears, Chairman of the Football Association, received a call from a soldier called Edward Bletchley. Bletchley offered to return the cup for a sum of £15,000 as long as nothing was said to the police. Mears agreed to pay the ransom, **but** kept the second part of the bargain too. **Consequently**, when Bletchley turned up at a secret location to collect the money, the police knew nothing about it. He was arrested and charged with theft, **although** he later claimed to be only a middle man who would receive just £500 for his trouble.

Bletchley soon found himself in prison but the cup was still missing. Then, a few days later, a man called David Corbett was taking his dog Pickles for a walk when the dog dragged him over to a corner of the garden. Under the hedge was a parcel wrapped in newspaper. David's first thought was that it might be a bomb **but** when he pulled off the newspaper, there was the world cup. **As a result of** his find, Pickles became an instant celebrity. **Because** he had saved the world cup, he was allowed to attend the players' banquet and finish up the scraps. **Moreover**, David and Pickles went on to make a number of television appearances.

Although Pickles has been dead for many years, David seldom thinks about him now and has all but forgotten that he once owned one of the most famous dogs in history.

Commas

1 Commas are used in place of a link word if it has moved to the beginning of the sentence.
2 Commas are used after or around some adverbial words and phrases.
3 Commas are used before some conjunctions.
4 Commas are used around short phrases which define or clarify the noun before them.

a) Mike Robinson, the famous film director, is currently holidaying in the Caribbean.
b) Most reality TV has no merit at all. It is, however, very popular with viewers.
c) Men used to be more reluctant to apologise, but this is changing.
d) Although I apologised to Susan, she still hasn't forgiven me.

Colons and semicolons

1 A .... is used in titles to indicate a subheading.
2 A .... can sometimes be used instead of a full stop or a link word to join two sentences.
3 A .... is used after a complete sentence which announces what is to come next.
4 A .... is sometimes used instead of a full stop before an adverb like *therefore.*

a) There are four castes in traditional Indian society: priests, warriors, businessmen and servants.
b) I remember James as a little boy; now he's just become our member of parliament.
c) I am reading *Universal Man: an Introduction to Anthropology* by Peter Rowe.
d) Modesty is usually considered a virtue; however, extreme modesty can prevent people from fulfilling their true potential.

THE ORIGINS OF FLIGHT

How some members of the animal kingdom developed the ability to fly — 1 ....
remains mysterious. There seem to be two possibilities the ground-up — 2 ....
theory and the trees-down theory. According to the first theory these — 3 ....
animals developed flight by first learning to leap into the air. As their — 4 ....
forelimbs became stronger they were able to leap higher and eventually — 5 ....
take off in short bursts. The other theory suggests that animals began — 6 ....
their path to flight by jumping from tree to tree or a tree to the ground. — 7 ....

One theory suggests that birds are descended from dinosaurs if so then — 8 ....
the ground-up theory is more likely as there seems no evidence that — 9 ....
dinosaurs lived in trees. But if we look at the case of bats the opposite — 10 ....
is true there is some evidence that they are related to squirrels. Despite — 11 ....
their obvious similarity it seems that bats and birds are unrelated and — 12 ....
indeed developed their ability to fly via two completely different paths. — 13 ....

Part one

First of all, I'd like to highlight just a few of the catastrophic decisions that the local council has made affecting people in this city since the last election.

In fact, 1 .... And now look at the results: a shortage of bus drivers, long queues at bus stops for passengers, to say nothing of totally inadequate maintenance that leads to breakdowns and cancellations. And those without cars have no alternative – is this how we encourage people to use public transport? And there are other issues. 2 .... We should all be appalled at the current situation where many of our older citizens are still having to pay for their bus passes whereas in some well run cities, pensioners have had free travel for years. This surely cannot be right. Council officials put this down to the growing number of pensioners but, quite frankly, 3 .... The reality, of course is that they have mismanaged the entire financial situation. 4 .... If we do, we have only ourselves to blame for the chaos that will surely follow.

A never have I heard such a lame excuse.

B no sooner had they got into power than they made cutbacks in the financial help given to public services, especially public transport.

C Under no circumstances can we allow this council to be re-elected.

D At no time in living memory have we had such poor concessions on public transport for the elderly.

Part two

No sooner …	Not only …
Under no circumstances …	Never before …
Not one/once …	At no time …
Only by …	Not until …

Worksheet 1

Everyone knows the story of Archimedes and how he was given the task of finding out the true gold content of the king's crown. For a long time, he was at a (1) .... to know how to do it. Then, one day when he was stepping into his bath, he noticed that some of the water overflowed onto the floor. Instantly, he was struck by the realisation that a crown made of pure gold would displace a different amount of water from one made of an alloy. (2) .... with excitement at having found the solution, he ran into the street shouting 'Eureka', or 'I have found it.' The parable illustrates the way in which new ideas or solutions to problems sometimes seem to come to us (3) ...., as a sudden flash of insight. Similar moments of (4) .... are claimed for Descartes' invention of co-ordinates and Crick and Watson's discovery of DNA.

Of course, the long and laborious processes of thought and logical (5) .... cannot be ignored. On the contrary, Eureka moments seem to occur only after a long period of consideration of the problem in hand. The interesting thing is that the most famous ones seem to have occurred when the person was doing something (6) .... with the problem in question, like having a bath. What seems to happen is that, after puzzling over the matter for some time, people sometimes feel that they have reached a mental block. The only way to progress is by synthesising what they already know with totally different information. At the same time, those many hours spent grappling with the problem have unconsciously put the brain on the (7) .... When the thinker is involved with something apparently (8) ...., the primed brain reaches out and makes a connection between this and the problem. To the thinker, the solution seems to have come like a flash of inspiration, out of nowhere.

1 **LOSE** 2 **COME** 3 **MIRACLE** 4 **REVEAL**
5 **DEDUCE** 6 **CONNECT** 7 **LOOK** 8 **RELATE**

Worksheet 2

Everyone knows the story of Archimedes and how he was given the task of finding out the true gold content of the king's crown. For a long time, he was at a loss to know how to do it. Then, one day when he was stepping into his bath, he noticed that some of the water (1) .... onto the floor. Instantly, he was struck by the (2) .... that a crown made of pure gold would displace a different amount of water from one made of an alloy. Overcome with excitement at having found the solution, he ran into the street shouting 'Eureka', or 'I have found it.' The parable illustrates the way in which new ideas or solutions to problems sometimes seem to come to us miraculously, as a sudden flash of (3) .... Similar moments of revelation are claimed for Descartes' invention of co-ordinates and Crick and Watson's discovery of DNA.

The existence of moments like this does not mean that the long and (4) .... processes of thought and logical deduction can be ignored. On the contrary, Eureka moments seem to occur only after a long period of (5) .... of the problem in hand. The interesting thing is that the most famous ones seem to have occurred when the person was doing something unconnected with the problem in question, like having a bath. What seems to happen is that, after puzzling over the matter for some time, people sometimes feel that they have reached a mental block. The only way to progress is by (6) .... what they already know with totally different information. At the same time, those many hours spent grappling with the problem have (7) .... put the brain on the lookout. When the thinker is involved with something apparently unrelated, the primed brain reaches out and makes a connection between this and the problem. To the thinker, the solution seems to have come like a flash of (8) ...., out of nowhere.

1 **FLOW** 2 **REALISE** 3 **SIGHT** 4 **LABOUR**
5 **CONSIDER** 6 **SYNTHESIS** 7 **CONSCIOUS** 8 **INSPIRE**

Worksheet 1

1 The Montreal Stadium

Montreal's Olympic stadium was built for the 1976 Olympic Games. The design was extremely ambitious, and featured a retractable roof, which could be closed up around a tall tower, like a huge umbrella.

It was originally forecast to cost 120 million Canadian dollars but by the time of the Olympics it had already cost 250 million. To make matters worse, it was far from finished because the tower and the famous roof were missing. The long-anticipated retractable roof did not even arrive from its origin in Paris until 1981 and then it sat idle for several more years until the city found the money to install it. It was then found that the retracting mechanism did not work properly.

In 1991, part of the roof's support gave way, causing a 55-ton slab of concrete to crash to the ground. Miraculously, no one was hurt. It was then decided to give up on the idea of the retractable roof and hopefully solve the problem once and for all by fixing on a permanent one. This project cost another 57 million dollars but the new roof was finally attached in 1998. Then in January 1999, a large part of this new roof fell in, due to the weight of the snow and ice on it.

The stadium is now closed during the winter months for safety reasons and it sits vacant for most of the rest of the year due to its inconvenient location. The total cost of the project, a massive 1.47 billion dollars, was finally paid off in 2006. The locals refer to it as 'the Big O' or 'the Big Mistake'.

2 The *Saro Princess*

The *Saro Princess* was a large flying boat, built by the British company Sanders-Roe Ltd.

Before the Second World War, it seemed obvious that journeys across the Atlantic would be made by flying boats. After all, it was reasoned, aeroplanes were not very reliable and you would need to be able to make an emergency landing on water. Accordingly, work on the *Saro Princess* began in 1946. It was intended to be a supreme sea plane that could serve the British Empire.

Unfortunately, people were not aware of just how rapidly the technology for flying would develop. The *Saro's* ambitious design also took much longer than anticipated to complete. The finished model was finally unveiled at an air show in 1952 but by that time many international airports were already operating and the design was already obsolete. In all, three models of the *Saro* were built but only one of them ever flew.

Sentences to complete

It was going to be ....

It was to be ....

It was hoped that it ....

People had no idea that ....

No one suspected that ....

People did not foresee that ....

In the event it was ....

As it turned out, the costs were to be ....

If .... had known ...., he/she/they ....

Worksheet 2

1 The Waterloo Vase

The Waterloo Vase is an enormous vase, five metres high and weighing about 20 tons.

When Napoleon was passing through Italy on his way to the Russian front, he was impressed by the enormous blocks of marble that had been hewn from the mountains. He ordered one of them to be saved so that it could be turned into a victory trophy for himself. Unfortunately, of course, his dreams of victory came to an end at the Battle of Waterloo. The marble was therefore offered to the king of Britain, George IV, instead. The king also liked the idea of having a war trophy and commissioned the sculptor, Richard Westmacott, to turn it into a massive vase. The hope was that this would become a widely admired work of art to celebrate Britain's victory over France.

The facts turned out rather differently. The vase was originally intended to stand in Windsor Castle, but it was so heavy that the floor could not bear its weight. It was decided to give it to the National Gallery instead. However, they did not want it either and eventually returned it to the monarchy, to King Edward VII in 1906. The vase was finally placed in a secluded area of the garden in Buckingham Palace, where it stands today.

2 The Sydney Cross City Tunnel

The Sydney Cross City Tunnel links Darling Harbour on the western edge of the city with the suburbs on the eastern side. It was first opened in August 2005 and the hope was that it would ease traffic congestion in the city centre.

Unfortunately, the original projection for the number of cars using the tunnel turned out to be wildly optimistic. It was forecast that 85,000 vehicles a day would drive through it but the real figure was only about 25,000. In an attempt to encourage motorists to use it, a toll-free period was declared. At the end of this period, the number of cars had increased to 53,000. The organisers then reinstated the charge and the figure promptly dropped again.

As it was so expensive, the tunnel failed in its intention to reduce congestion because scores of motorists started to drive through the back streets of Sydney to avoid paying the toll. To make matters worse, a number of roads had been closed due to the construction of the tunnel, which made the gridlock even worse. Eventually, the government was forced to reopen some of the closed roads at further expense.

With debts of over 500 million Australian dollars, the tunnel has now gone into receivership. Who will finally foot the bill is uncertain.

Sentences to complete

It was going to be ....

It was to be ....

It was hoped that it ....

People had no idea that ....

No one suspected that ....

People did not foresee that ....

In the event it was ....

As it turned out, the costs were to be ....

If .... had known ...., he/she/they ....

Worksheet 1

deep	look	strike	rich	head	trial

1 The area was very suitable for agriculture as it had high rainfall and .... soil.

2 I found him staring out of the window, ... in thought.

3 As no one showed me how to use the computer programme, I had to learn by .... and error.

4 She has a reputation for keeping a cool .... in a crisis.

5 Oranges and grapefruits are both .... in vitamin C.

6 It can't be midday yet because I've just heard your clock .... eleven.

7 I think you need to take a long hard .... at the effects that your behaviour is having.

8 Her singing voice is surprisingly .... for a woman.

9 It's good that we have such happy times to .... back on.

10 The leader of the rebellion was put on .... and later executed.

11 The country is planning a mass protest, beginning with a general .... today.

12 Last week, the prime minister flew to Russia for talks with the new .... of state.

13 After his attempt to overrule parliament, the king found himself in .... trouble.

14 As he was walking through the grass, he felt his foot .... against something hard.

15 That's a really upmarket coffee shop where all the .... and famous go.

16 By the end of the day, the army decided to retreat and .... for the hills.

17 As he finally entered the room, she gave him an angry ....

18 The system was introduced on a .... basis and was to be reviewed after one month.

Worksheet 1 answers

1 rich 2 deep 3 trial 4 head 5 rich 6 strike 7 look 8 deep 9 look
10 trial 11 strike 12 head 13 deep 14 strike 15 rich 16 head 17 look 18 trial

Worksheet 2

rule	cut	fire	single	play	mark

1 We cannot .... out the possibility that the king was murdered by his younger son.

2 The government have come under .... for their financial mismanagement.

3 For a healthier diet, you need to .... down on sugar and fat.

4 After the coronation, the queen went on to .... for over 40 years.

5 I can't get this dirty .... off my shirt collar.

6 The rioters smashed several windows and set .... to a number of parked cars.

7 *Love's Labour's Lost* is certainly an early work, even if it is not the first .... that Shakespeare wrote.

8 It is difficult to bring up children as a .... parent.

9 The death of Elizabeth I is often said to .... the beginning of a new era.

10 When I was young, I didn't like school and I often used to .... truant.

11 The chancellor tried to .... down the significance of the rise in inflation.

12 How can there have been so many people in the street and not a .... witness to the crime?

13 In their election manifesto, the party promised to .... the basic rate of tax.

14 There is an unwritten .... that staff do not do online shopping during work hours.

15 The prince was a weak and moody young man who was not .... out for the responsibilities of kingship.

16 At school, the games master used to .... me out for criticism, just because I was overweight.

17 The police were obviously ready to .... on the crowd if there was any trouble.

18 A great many students have asked me what the pass .... is for the exam.

Worksheet 2 answers

1 rule **2** fire **3** cut **4** rule **5** mark **6** fire **7** play **8** single **9** mark
10 play **11** play **12** single **13** cut **14** rule **15** cut **16** single **17** fire **18** mark

Worksheet 1

Rules

1 For words ending with a consonant plus *y*, change *y* to *i* before adding any suffix except *-ing*.

2 For words ending in a single vowel plus a single consonant, double the final consonant before adding a suffix if the final syllable is stressed.

3 For words ending in *e*, drop the *e* before adding a suffix beginning with a vowel …

4 … but keep the *e* if the suffix begins with a consonant.

5 When a prefix ends with the same letter as the first one of the word, keep both letters.

6 The *-ful* suffix at the end of many adjectives is always written with one *l* …

7 … but if you add the *-ly* adverb suffix, then there will be two *l*s.

8 *i* before *e*, except after *c*, when the pronunciation is /iː/ …

9 … but with other pronunciations (usually /eɪ/), the *e* comes first.

10 When the *c* is pronounced 'sh', the *i* comes first.

Examples

A believe, receive, ceiling, field

B wonderful, useful, harmful, peaceful

C carried, parties, happier, worrying

D ancient, species, efficient, sufficient

E stopped, hottest, preferred, mattered

F improvement, careful, definitely, advertisement

G hopefully, carefully, unhelpfully, beautifully

H unnatural, dissatisfied, immoral, irresistible

I weight, freight, height, sleigh

J advisable, famous, creative, driving

Worksheet 2

	correction	exception or rule?
1 brief begining permitted happiness		
2 nieghbour happened relief hateful		
3 necessarily shield fortunatly deficient		
4 incurable benefited achieve protien		
5 occured unnecessary thoughtfully awful		
6 measurement niece dissappear closure		
7 mispelt conceited unimaginable chief		
8 iresponsible perceive lately immaterial		
9 entirely retrieve studing writing		
10 unintentional admitted leisure successfull		
11 couragous adequately judgement belief		
12 arguement interrelated adventurous safely		

Certificate in Advanced English quiz

Try this quiz to see how much you know about the exam.

1 Is there any extra time for transferring answers to the mark sheets? ....
2 Do I have to write in pen or pencil on the mark sheets? ....
3 What happens if I shade in two lozenges on the mark sheets? ....
4 Does spelling have to be correct? ....
5 Is there any negative marking in multiple-choice questions? ....
6 Do I have to pass each paper to pass the exam? ....
7 How soon do I get my results? ....

Reading

8 How many parts are there in the paper? ....
9 How long do I have to complete the paper? ....
10 What should I look at first, the text or the questions? ....

Writing

11 How many questions do I have to answer? ....
12 Can I write answers in pencil? ....
13 Can I use correction fluid? ....
14 Is it a good idea to write a rough copy and then copy it out neatly? ....
15 Is it important to write in paragraphs? ....
16 Does handwriting count? ....
17 Is it a good idea to write on alternate lines? ....
18 What happens if my answers are too short or too long? ....
19 Should I spend equal time on each answer? ....

Use of English

20 How many parts are there in the paper? ....
21 What happens if I write two possibilities for one gap in the cloze passage? ....
22 Does the key word always have to be altered in the word formation exercise? ....
23 What happens if I write more than six words in the key word transformations? ....

Listening

24 How many parts are there in the paper? ....
25 How many times do I hear each passage? ....
26 In Part 2, do I write the words I hear or should I paraphrase? ....

Speaking

27 Can I ask the interlocutor to repeat his/her instructions? ....
28 Will I lose marks if I do not talk for one minute in Part two? ....
29 What happens if my partner is too quiet or too talkative? ....
30 What should I do if I don't know a word? ....

Reading

UNIVERSITY of CAMBRIDGE
ESOL Examinations

Do not write in this box

Candidate Name
If not already printed, write name in CAPITALS and complete the Candidate No. grid (in pencil).

Candidate Signature

Examination Title

Centre

Supervisor:
If the candidate is ABSENT or has WITHDRAWN shade here

Centre No.

Candidate No.

Examination Details

0	0	0	0
1	1	1	1
2	2	2	2
3	3	3	3
4	4	4	4
5	5	5	5
6	6	6	6
7	7	7	7
8	8	8	8
9	9	9	9

Candidate Answer Sheet

Instructions

Use a **PENCIL** (B or HB).

Mark ONE letter for each question.

For example, if you think B is the right answer to the question, mark your answer sheet like this:

0	A	B	C	D	E	F	G	H

Rub out any answer you wish to change using an eraser.

1	A	B	C	D	E	F	G	H
2	A	B	C	D	E	F	G	H
3	A	B	C	D	E	F	G	H
4	A	B	C	D	E	F	G	H
5	A	B	C	D	E	F	G	H
6	A	B	C	D	E	F	G	H
7	A	B	C	D	E	F	G	H
8	A	B	C	D	E	F	G	H
9	A	B	C	D	E	F	G	H
10	A	B	C	D	E	F	G	H
11	A	B	C	D	E	F	G	H
12	A	B	C	D	E	F	G	H
13	A	B	C	D	E	F	G	H
14	A	B	C	D	E	F	G	H
15	A	B	C	D	E	F	G	H
16	A	B	C	D	E	F	G	H
17	A	B	C	D	E	F	G	H
18	A	B	C	D	E	F	G	H
19	A	B	C	D	E	F	G	H
20	A	B	C	D	E	F	G	H

21	A	B	C	D	E	F	G	H
22	A	B	C	D	E	F	G	H
23	A	B	C	D	E	F	G	H
24	A	B	C	D	E	F	G	H
25	A	B	C	D	E	F	G	H
26	A	B	C	D	E	F	G	H
27	A	B	C	D	E	F	G	H
28	A	B	C	D	E	F	G	H
29	A	B	C	D	E	F	G	H
30	A	B	C	D	E	F	G	H
31	A	B	C	D	E	F	G	H
32	A	B	C	D	E	F	G	H
33	A	B	C	D	E	F	G	H
34	A	B	C	D	E	F	G	H
35	A	B	C	D	E	F	G	H
36	A	B	C	D	E	F	G	H
37	A	B	C	D	E	F	G	H
38	A	B	C	D	E	F	G	H
39	A	B	C	D	E	F	G	H
40	A	B	C	D	E	F	G	H

A-H 40 CAS

DP594/300

Listening

Part 1			
1	A	B	C
2	A	B	C
3	A	B	C
4	A	B	C
5	A	B	C
6	A	B	C

Part 2 (Remember to write in CAPITAL LETTERS or numbers)		Do not write below here
7		7 1 0 u
8		8 1 0 u
9		9 1 0 u
10		10 1 0 u
11		11 1 0 u
12		12 1 0 u
13		13 1 0 u
14		14 1 0 u

Part 3				
15	A	B	C	D
16	A	B	C	D
17	A	B	C	D
18	A	B	C	D
19	A	B	C	D
20	A	B	C	D

Part 4								
21	A	B	C	D	E	F	G	H
22	A	B	C	D	E	F	G	H
23	A	B	C	D	E	F	G	H
24	A	B	C	D	E	F	G	H
25	A	B	C	D	E	F	G	H
26	A	B	C	D	E	F	G	H
27	A	B	C	D	E	F	G	H
28	A	B	C	D	E	F	G	H
29	A	B	C	D	E	F	G	H
30	A	B	C	D	E	F	G	H

Use of English

UNIVERSITY of CAMBRIDGE
ESOL Examinations

Do not write in this box

Candidate Name
If not already printed, write name in CAPITALS and complete the Candidate No. grid (in pencil).

Candidate Signature

Examination Title

Centre

Supervisor:
If the candidate is ABSENT or has WITHDRAWN shade here

Centre No.

Candidate No.

Examination Details

0	0	0	0
1	1	1	1
2	2	2	2
3	3	3	3
4	4	4	4
5	5	5	5
6	6	6	6
7	7	7	7
8	8	8	8
9	9	9	9

Instructions
Use a PENCIL (B or HB).
Rub out any answer you wish to change.

Part 1: Mark ONE letter for each question.
For example, if you think B is the right answer to the question, mark your answer sheet like this: 0 A B C D

Parts 2, 3, 4 and 5: Write your answer clearly in CAPITAL LETTERS.
For Parts 2, 3 and 4, write one letter in each box. 0 EXAMPLE

Candidate Answer Sheet

Part 1

1	A	B	C	D
2	A	B	C	D
3	A	B	C	D
4	A	B	C	D
5	A	B	C	D
6	A	B	C	D
7	A	B	C	D
8	A	B	C	D
9	A	B	C	D
10	A	B	C	D
11	A	B	C	D
12	A	B	C	D

Part 2

Do not write below here

13		13 1 0 u
14		14 1 0 u
15		15 1 0 u
16		16 1 0 u
17		17 1 0 u
18		18 1 0 u
19		19 1 0 u
20		20 1 0 u
21		21 1 0 u
22		22 1 0 u
23		23 1 0 u
24		24 1 0 u
25		25 1 0 u
26		26 1 0 u
27		27 1 0 u

Continues over →

CAE UoE

DP597/301

Part 3

Do not write below here

28		28 1 0 u
29		29 1 0 u
30		30 1 0 u
31		31 1 0 u
32		32 1 0 u
33		33 1 0 u
34		34 1 0 u
35		35 1 0 u
36		36 1 0 u
37		37 1 0 u

Part 4

Do not write below here

38		38 1 0 u
39		39 1 0 u
40		40 1 0 u
41		41 1 0 u
42		42 1 0 u

Part 5

Do not write below here

43		43 2 1 0 u
44		44 2 1 0 u
45		45 2 1 0 u
46		46 2 1 0 u
47		47 2 1 0 u
48		48 2 1 0 u
49		49 2 1 0 u
50		50 2 1 0 u

denote Print Limited 0121 520 5100